# Precision Agriculture - Emerging Technologies

*Edited by Redmond R. Shamshiri, Sanaz Shafian and Ibrahim A. Hameed*

Published in London, United Kingdom

Precision Agriculture - Emerging Technologies
http://dx.doi.org/10.5772/intechopen.111335
Edited by Redmond R. Shamshiri, Sanaz Shafian and Ibrahim A. Hameed

Contributors
Ahmad Kamil Zolkafli, Anielle Christine A. Almeida Silva, Anna V.B. Borges, Arina Mohd Noh, Badril Abu Bakar, Beom-Sun Kang, Christopher Boon Sung Teh, Cong-Chuan Pham, Eliete A. Alvin, Ho-Song An, Ibrahim A. Hameed, Jaesung Park, Joao M. P. R. Serrano, José Marques Da Silva, Jusnaini Muslimin, Karthick T., Luis Galeano-Vasco, Manuela Ortega Monsalve, Marcos Vinicius Da Silva, Mario Cerón-Muñoz, Marisol Medina-Sierra, Maryam Behjati, Mehmet Ali Çullu, Mehmet Hadi Suzer, Mehmet Şenbayram, Mohamad Shukri Zainal Abidin, Mohammed El Idrissi, Mohd Nizam Zubir, Muhammad Hariz Musa, Nilvanira D. Tebaldi, Pradeep Kumar Garg Garg, Prince Kumar, Ramlan Ismail, Redmond R. Shamshiri, Rodrigo C. Rosa, Ryuta Kojo, Shakib Shahidian, Siti Noor Aliah Baharom, Siva K. Balasundram, Sumaiya Islam, Sun-Ok Chung, Tinah Manduna Mutabazi, Valentina Parilli-Ocampo, Vasudevan N., W.A. Balogun, Wei-Chih Lin, Wesley S. M. Ribeiro



First published in London, United Kingdom, 2024 by IntechOpen
IntechOpen is the global imprint of INTECHOPEN LIMITED, registered in England and Wales, registration number: 11086078, 167-169 Great Portland Street, London, W1W 5PF, United Kingdom

British Library Cataloguing-in-Publication Data
A catalogue record for this book is available from the British Library

Additional hard and PDF copies can be obtained from orders@intechopen.com

Precision Agriculture - Emerging Technologies
Edited by Redmond R. Shamshiri, Sanaz Shafian and Ibrahim A. Hameed
p. cm.

This title is part of the Agricultural Sciences Book Series, Volume 11
Topic: Agronomy and Horticulture
Series Editor: W. James Grichar
Topic Editor: Ibrahim Kahramanoglu
Associate Topic Editors: Murat Helvaci and Olga Panfilova

Print ISBN 978-0-85014-279-2
Online ISBN 978-0-85014-280-8
eBook (PDF) ISBN 978-0-85014-281-5
ISSN 3029-052X

For EU product safety concerns :
IN TECH d . o . o . , Prolaz Marije Krucifikse Kozulić 3 , 51000 Rijeka , Croatia ,
info@intechopen . com or visit our website at intechopen . com .

IntechOpen Book Series

# Agricultural Sciences

Volume 11

## Aims and Scope of the Series

The importance of agriculture cannot be overstated. It helps sustain life, as it gives us the food we need to survive and provides opportunities for economic well-being. Agriculture helps people prosper around the world and combines the creativity, imagination, and skill involved in planting crops and raising animals with modern production methods and new technologies. This series includes two main topics: Agronomy and Horticulture, and Animal Farming. This series will help readers better understand the intricacies of production agriculture and provide the new knowledge that is required to be successful. The success of a farmer in modern agriculture requires knowledge of events happening locally as well as globally that impact input decisions and ultimately determine net profit.

# Meet the Series Editor

W. James Grichar has been employed with Texas A&M AgriLife Research for over 45 years with an emphasis on research in agronomy, plant pathology, and weed science. He obtained his BS from Texas A&M in 1972 and his Masters of Plant Protection in 1975. He has published 195 journal articles, over 330 research reports and briefs, 11 book chapters, and over 300 abstracts of profession meetings. He also directs research in many crops including corn, grain sorghum, peanuts, and sesame. He has held various positions in different professional societies including the American Peanut Research and Education Society, Southern Weed Science Society, and Texas Plant Protection Conference in addition to being Associate Editor for Peanut Science and Weed Technology. Significant accomplishments have included spearheading efforts to determine the optimum planting time for soybean production along the upper Texas Gulf Coast. These efforts have shown growers that soybean yields can be improved by 10 to 20% by following a late March to early April plant date. He also has been instrumental in developing a herbicide program for peanut production in the south Texas growing region. Through the development and use of herbicides that are effective against major weed problems in the south Texas region, peanut yields have increased by 25 to 30%.

# Meet the Volume Editors

Dr. Redmond R. Shamshiri is a scientist at the Leibniz-Institut für Agrartechnik und Bioökonomie working on agro-mechatronics projects. He holds a Ph.D. in automation and control with a focus on crop production systems. His research areas include the Internet of Robotic Things for the teleoperation of agricultural mobile platforms, distributed collision avoidance systems for assisted autonomous navigation, LPWAN sensors for digital farming applications, and optimization of controlled environment crop production systems. He is currently working on the evaluation of 5G technology to reduce yield losses from potato fields.

Dr. Sanaz Shafian is an assistant professor of smart farming at Virginia Tech University. Prior to this, she was assistant professor at the University of Idaho. Her expertise lies in remote sensing research, with a focus on using advanced technologies and methodologies for economically and environmentally sustainable crop and tree monitoring and management. She integrates satellite/drone images and AI to develop methodologies for environmental monitoring, crop modeling, and water and nutrient conservation, and she has published widely on these topics. She has been involved in several USDA projects. With the University of Idaho, she led an educational and outreach project to initiate the Precision Agriculture certificate. Sanaz is currently working on two agriculture-related projects, one mapping invasive plants in temperate forests and evaluating forest health in Virginia and another looking at developing smart tools for early detection of soybean diseases.

Dr. Ibrahim A. Hameed holds a Ph.D. in AI from Korea University, South Korea, and a Ph.D. in field robotics from Aarhus University, Denmark. He is currently a professor at the Norwegian University of Science and Technology (NTNU) and the Norwegian University of Life Science (NMBU), Norway. Hameed is an IEEE senior member, elected chair of the IEEE Computational Intelligence Society (CIS) Norway section, and founder and head of the Social Robots Lab in Ålesund. He has participated in several research projects funded by both Norwegian and European research councils. His current research interests include artificial intelligence, machine learning, optimization, and robotics.

# Contents

# Preface

Precision agriculture is experiencing rapid technological advancements through the use of cutting-edge tools and methods to optimize every aspect of crop production and management, with the ultimate goal of increasing efficiency, sustainability, and productivity. This book is a comprehensive exploration of the latest innovations driving the field of precision agriculture forward. Our objective is to provide a detailed look at how emerging technologies like aerial imaging, wireless sensing, and mobile apps, as shown in **Figure 1**, are reshaping agricultural practices worldwide. By delving into other advancements such as the Internet of Things (IoT), digital twins, blockchain, and robotics, we aim to highlight not only the significant benefits they bring, such as improved efficiency, reduced labor costs, and enhanced safety, but also the challenges they present, including high adoption costs, regulatory hurdles, and the need for specialized skills.

The most notable emerging technologies in precision agriculture, in our opinion, are those that enhance data connectivity, processing, and sharing. One of the most revolutionary advances in this field is the introduction of fifth-generation mobile networks (5G), which are reshaping and redefining precision agriculture by opening up new possibilities for distributed automation systems, including remote control of collaborative robots and groups of small, unmanned machinery. These systems can independently carry out specific operations like weeding and spraying, thanks to IoT-based cloud computing services. While similar solutions have been tested or

**Figure 1.**
*Schematic demonstration of new technologies, including wireless sensing, aerial imaging, and mobile apps, for site-specific monitoring of field parameters in modern precision agriculture. (Picture courtesy of Adaptive AgroTech, designed by Keywan Majidi)*

put into commercial use, ensuring stable and secure connections between different components remains a persistent concern. Concerning scenario analysis for open-field and closed-field crop production systems, the relatively new groundbreaking technology called digital twins enables real-time evaluation, optimization, and predictive control of complex systemic processes. This interaction, taking place within a centralized or cloud-based architecture, allows for process optimization, updates to control parameters, and the generation of predictive solutions for hypothetical scenarios. This capability has a good potential for enhancing agricultural productivity and efficiency. Blockchain technology also holds transformative potential for the agricultural sector by creating decentralized and secure networks that contribute to better transparency, traceability, and efficiency in the agricultural supply chain. Blockchain can be used to create a digital ledger that records all data generated by sensors and controllers, facilitating informed decisions about planting, fertilizing, and harvesting crops.

The chapters in this book represent the culmination of extensive research and practical insights from academics and experts in the field. Each chapter provides detailed research reports that explore new methods and perspectives in precision agriculture applications, giving readers an overview of both the theoretical foundations and practical uses of some of the most recent innovative solutions. From fundamental research to applied practices, the content is designed to cater to two audiences: the academic community and practitioners in the agricultural sector. For academics, this book serves can serve as a resource to understand the current state of precision agriculture technologies and their potential for future development. For farmers and agricultural professionals, it offers practical guidance on how to make the most of these technologies to improve farm management.

During the editorial phase of this book, we received significant support from our colleague, Ms. Maryam Behjati of Adaptive AgroTech Consultancy. We would like to acknowledge her contributions and express our deep appreciation.

**Redmond R. Shamshiri**
Leibniz Institute for Agricultural Engineering and Bioeconomy,
Potsdam, Germany

**Sanaz Shafian**
Virginia Tech,
United States of America

**Ibrahim A. Hameed**
Norwegian University of Science and Technology,
Norway

Chapter 1

# Agricultural Mobile Robots for Plant Health Assessment and Drought Stress Detection

*Maryam Behjati, Redmond R. Shamshiri and Ibrahim A. Hameed*

## Abstract

The vulnerability of plants to various threats, such as insects, pathogens, and weeds, poses a significant risk to food security, particularly before harvest. Mobile robots are used in digital agriculture as a breakthrough approach to address challenges in crop production, such as plant health assessment and drought stress detection. This chapter aims to explore the application of agricultural mobile robots equipped with advanced sensing technologies and computer vision algorithms, along with their key features, to enhance crop management practices. An overview of some the platforms with different steering mechanisms, sensors, interfaces, communication, and machine learning has been provided along with case studies on the use of robots for collecting data on plant health indicators such as physiological parameters, leaf coloration, and soil moisture levels. Recent trends in this area show that by utilizing machine learning techniques such as convolutional neural networks (CNNs) and support vector machines (SVMs), the collected data are analyzed to identify symptoms of plant diseases, nutrient deficiencies, and drought stress, facilitating timely interventions to mitigate crop losses. The integration of Internet of robotic things into existing practices are also discussed with respect to cost-effectiveness, scalability, and user acceptance.

**Keywords:** agricultural robotics, modern precision agriculture, plant health, digital farming, mobile robots

## 1. Introduction

Plant health is a critical aspect of agriculture-related to achieving farming goals, and overcoming challenges is plant health. The complex interactions between the plant hosts, various plant pathogens, and different environmental factors influence plant health [1]. In addition, different products face different types of biotic and abiotic stresses. Abiotic stresses including drought, temperature fluctuations, high soil salinity, and oxidative stress. These stresses can cause permanent damage to the plants, such as stopping growth, disrupting metabolism, reducing performance, and changing genetic behavior [2]. Drought occurs over a long period due to the insufficient availability of fresh water, but with the increase in global temperature and climate change, drought has intensified.

Drought, in addition to affecting food production, reduces the level of underground water, increases soil erosion, and eventually can lead to other disasters such as fire, flood, and the spread of diseases [3]. Drought is classified into four separate categories: 1. Meteorological drought that occurs in areas with dry climate. 2. Hydrological drought that occurs in water shortage conditions, especially in surface and underground water levels, and is encountered after a few months of meteorological drought. 3. Agricultural drought is often associated with a decrease in the water level in the soil and, as a result, crop failure, which poses a threat to food supply around the world. 4. Socio-economic drought which is related to the failure of supply and demand of various goods due to drought [2]. If the drought in the plant can be detected in the early stages, these short-term responses will not cause much damage to the plants. However, if the plant is under stress for a long period, more serious and permanent damage is caused to the plant [4]. Traditional methods for plant health assessment typically involve manual and destructive techniques. These methods are usually laborious, time-consuming, and expensive. Moreover, the manual sampling and analysis procedures typically entail numerous steps that rely on human intervention and low precision [5].

To overcome these challenges, the adoption of new agricultural technologies and techniques is necessary. Precision agriculture emerges as a strategic approach that leverages electronic information and various technologies for data collection, processing, and analysis [6]. Modern agriculture has various important aspects, such as Unmanned Ground Vehicles (UGV) and Unmanned Aerial Vehicles (UAV). While drones possess the capability to measure large areas in a short period, they have drawbacks such as depending on weather conditions, sensor limitations, and data quality [7]. To overcome these limitations, UGV proves to be suitable. The development of UGV is divided into two primary approaches, including autonomous conventional vehicles and independent mobile robots. Autonomous conventional vehicles can modify and deploy existing farm machinery for automated agricultural tasks, including planting, spraying, fertilizing, and harvesting. Although these vehicles are highly reliable and perform well on uneven terrains, they may lack some adaptability features required for specific tasks, such as integrating sensors and working with varying row spacing during experiments. Since farm vehicles are designed for transporting people and towing heavy equipment, they are typically heavier and larger, which can lead to soil compaction and the need for robust steering mechanisms [8]. In contrast to the UAV approach, independently mobile robots are specifically designed to be lightweight and adaptable to different wheel spacings and sensor configurations. This enables them to be used for a wide range of experiments and terrains while minimizing soil compaction [9]. The integration of robots represents a revolutionary development in precision agriculture in recent years. The increasing prevalence of robots in agriculture is evident, with a diverse array of robotic solutions currently available in the market. The widespread use of agricultural robots has led to a reduction in the need for labor and an increase in the production of agricultural products [10]. The development of robotics integrated with remote sensing technology offers an opportunity to assess complex traits more effectively.

The objective of this chapter is to provide an overview of robotics in plant health assessment and its features, including steering, sensing, communication, and the application of machine vision and deep learning in robots.

## 2. Mobile robot platforms

Unmanned ground vehicles, particularly robots, are widely used in precision agriculture due to numerous advantages, including reduced reliance on human resources and higher accuracy. The mobile platform is an essential component for the performance of any agricultural mobile robot. It plays a significant role in determining the overall system's effectiveness. One of the most important locomotion systems in robots is wheels and their arrangement together. The wheel system will play an important role in transportation, logistics, and distribution, and selecting the appropriate wheel type is crucial for kinematics and dynamics modeling. It is important to know that no universally optimal drive system can optimize maneuverability, controllability, and stability at the same time, however, the best type should be considered in the design of the robot according to the environmental conditions and the specific robot application.

### 2.1 Differential steering

The most prevalent arrangement of wheels in agrobots is differential steering. In a differential-drive mobile robot, the actuators connected to the wheels must be driven along the same velocity profile [11]. Some of the differential steering robots are shown in **Figure 1**, including (a) Xf-Rovim [13], and (b) RobHortic [12]. The Xf-Rovim, designed for early detection of *Xylella fastidiosa* (*X. fastidiosa*) in olive orchards, employs two DC motor control drivers, specifically H-Bridge modules. These modules enable the operator to have precise control over the speed, torque, and direction of rotation for each wheel. The utilization of a dedicated driver for each wheel enhances maneuverability. This feature is particularly useful for reducing the rotation angle and achieving greater precision in the robot's movements. RobHortic is a field robot designed to detect the presence of pests and diseases in horticultural crops. The robot contained a frame with four fat-bike wheels designed to navigate through uneven terrains seamlessly. While the front wheels remain fixed, the two rear wheels, comparatively smaller, can rotate freely, enhancing the robot's adaptability to diverse field conditions.

### 2.2 Skid steering

The Skid Steering mechanism operates like a differential wheel drive by controlling the relative speed of left and right drives. Its main advantages are mechanical simplicity, strength, and better maneuverability in straight movement. However, drawbacks include increased wheel wear due to skidding, reduced lifespan, and challenges in maintaining a straight path [14]. **Figure 2** shows examples of skid steering robotic platforms including (a) Robotianist [16], (b) Vinobot [17], and, (c) MARIA [15]. The Robotianist is an

**Figure 1.**
*Examples of differential steering robots, (a) Xf-Rovim, (b) RobHortic [12, 13].*

**Figure 2.**
*Examples of skid steering robots, (a) Robotianist, (b) Vinobot, and (c) MARIA [15–17].*

autonomous ground platform equipped with skid steering, designed for movement within the undergrowth of row crops like sorghum or corn. The power for each of the four pneumatic tires is supplied by a 200-watt brushless DC (BLDC) motor. This robotic system features a SCARA-style arm configuration, specifically tailored for measuring plant stem strength and collecting phenotypic data. Vinobot is a ground-based robotic platform designed for the assessment of plant phenotype, plant height, and the estimation of leaf area index. The robot is equipped with a linear slide guiding a JACO2 robotic arm from Kinova, enhancing lateral reach and enabling the handling of multiple sensors. MARIA is an indoor and outdoor robot developed using the Robot Operating System (ROS). MARIA's skid steering system, composed of off-the-shelf components, allows it to navigate through crop rows or greenhouses while conducting phenotyping and soil measurements.

### 2.3 Ackerman steering

Ackerman robots integrate a cutting-edge rotating mechanism to prevent sliding on curved paths. This involves aligning the rotation centers of all wheels to a singular point. As the rear wheels remain stationary, the pivotal point is strategically located along a line extending from the rear axle. Additionally, the left front wheel is programmed to turn at a greater angle than the outer front wheel, ensuring compliance with this fundamental principle [18]. The advantages of this design are increased control, better stability, and improved maneuverability on roads. **Figure 3** shows two different Ackerman robot platforms. AgriRover-01 is a mobile agricultural robot for collecting plant height information and estimating row spacing in corn plants. This robot is equipped with six motors: four in-wheel motors for driving and two motors for steering. The four driving wheels are divided into two pairs, front, and rear, with each pair employing a gear mechanism to enable simultaneous rotation. This configuration is implemented to execute Ackerman steering for effective and coordinated navigation in the agricultural environment [20]. Another example of Ackerman steering is DairyBioBot which is a fully automated phenotyping platform specifically used for Ryegrass biomass estimation. It has a lightweight four-wheel drive system that employs front wheel drive to ensure efficient mobility and precise data collection in field operations [19].

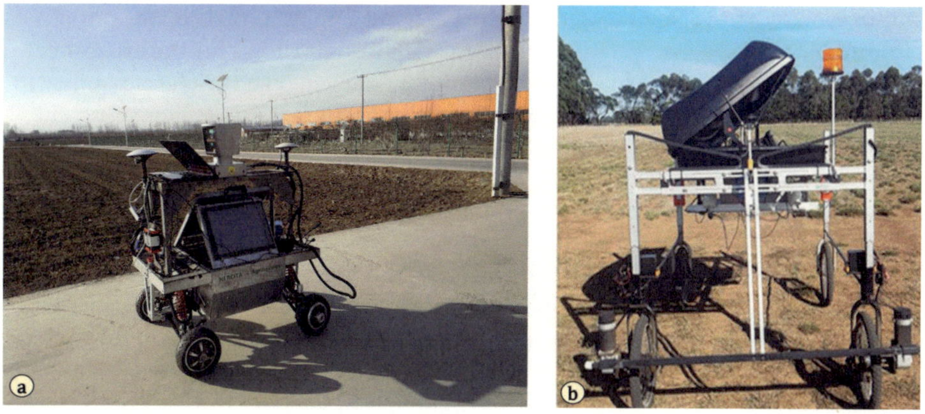


**Figure 3.**
*Examples of Ackerman steering robots, (a) AgriRover-01, (b) DairyBioBot [19, 20].*

## 2.4 Articulated steering

Articulated steering is like Ackerman steering in which the robot is propelled by deforming the entire chassis to achieve rotation. However, the mechanical complexity of articulated steering robots is higher compared to Ackerman robots. The wheel of an articulated steering is divided into two halves, front and rear, which are connected by a vertical hinge. To change the angle of the front part of the chassis and induce the desired rotation, a motor is employed [21]. Some of the robots that have articulated steering include PhenoBot 3.0 and Agri.Q are shown in **Figure 4**. PhenoBot 3.0 is a mobile robot designed for data acquisition in corn and sorghum plants within under-canopy navigation. This robot consists of a rover body in two parts, front and back, which are connected by an articulated steering joint. It is equipped with a central articulated steering mechanism and a differential gear-based power transmission that enables the robot to move quickly between narrow product rows with high efficiency [23]. Agri.Q is an eight-wheel articulated robot designed for precision agriculture applications, particularly crop sampling and monitoring in unstructured agricultural environments. The robot features a unique

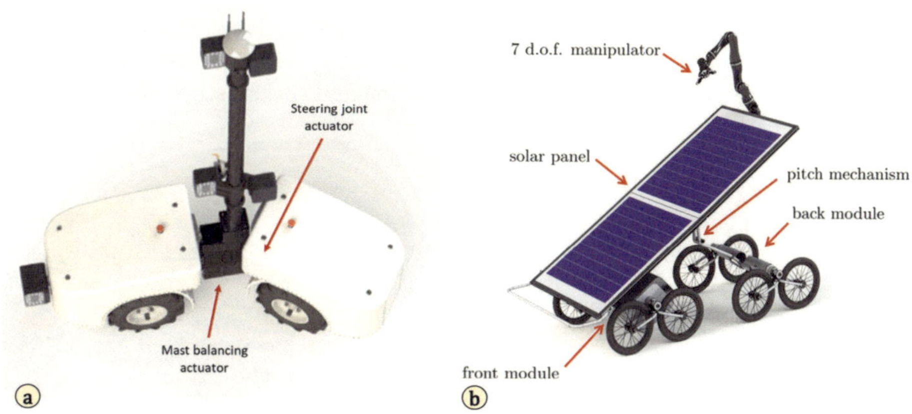


**Figure 4.**
*Examples of articulated steering robots, (a) PhenoBot 3.0, (b) Agri.Q [22, 23].*

combination of a novel articulated mobile base and a commercial collaborative manipulator with seven degrees of freedom. This design allows for the assurance of wheel contact with the ground, even on highly uneven terrain. The capability to distribute the robot's weight evenly among multiple contact points helps prevent excessive soil compaction during its operations [22].

### 2.5 Other steering mechanisms

Due to the diverse types of agricultural tasks and the complex conditions of the field, agricultural robots have various steering to accomplish their operational tasks. A weeding robot has been developed to address the challenges of using mobile robots to remove weeds in paddy fields. The robot churned up the soil to remove weeds and prevented further weed growth by blocking sunlight. The robot features two wheels, and its rotational speed is regulated by pulse width modulation (PWM) signals [24]. BoniRob is an agricultural mobile robot featuring Omnidirectional steering that incorporates diverse hardware modules according to research applications. These modules are strategically integrated to facilitate plant phenotyping, soil monitoring, weed control, and web-based communication. With its four driving and auto-steering wheels, BoniRob boasts a high level of maneuverability [25]. TERRA-MEPP is a tracked robot specifically designed for sorghum phenotype analysis that enables adaptable sensor positioning to address varying canopy heights during the growth cycle [26]. **Figure 5** displays some of the robots that use a different type of steering mechanism.

## 3. Mobile robot sensing system

In mobile robots, sensors are used for various purposes, such as phenotypic sensors for measuring phenotypic traits and perception sensors for navigation. Perception sensors play a crucial role in determining location and planning routes [14]. The phenotyping sensors include vision sensors and non-vision sensors. Optical sensors detect light in the 400–700 nm wavelength spectrum and provide red, green, and blue (RGB) reflectance values. In fact, sensors with different spectral bands enjoy different advantages. The two current technologies for vision-based sensors are CCD and CMOS [7]. The most widely used non-vision sensors are the hyperspectral camera, multispectral camera, thermal camera, and LiDAR sensor. LiDAR stands out as a highly employed sensor in robotics due to its reduced sensitivity to ambient light and its capacity to measure distances accurately without direct contact. Hyperspectral imaging is a combination of imaging and spectroscopy that produces

**Figure 5.**
*Examples of articulated steering robots are, (a) weeding robots, (b) BoniRob, and (c) TERRA-MEPP [24–26].*

multidimensional data. These sensors are capable of imaging with high resolution and narrow spatial range, so they can distinguish plant responses to stress. Unlike hyperspectral imaging, which uses hundreds of individual wavelengths or narrow wave bands to generate data, multispectral techniques use data from a range of wavelengths. In addition, multispectral imaging, compared with the hyperspectral method, is more affordable and more flexible. Thermal cameras detect radiation in the infrared wavelength range, with the resulting measurements being displayed where the pixels contain the temperature values [27].

### 3.1 Plant health assessment sensors

LiDAR sensors are increasingly applied in assessing crop health as they are able to measure 3D point clouds of crops [7]. While these sensors offer additional information such as depth and temperature, they have a higher financial cost, hindering their practical implementation [28]. To select the correct sensor for the agriculture robot, emphasizing efficiency rather than cost considerations. The agricultural environment introduces variables like temperature, humidity, and dust, which directly impact the performance of sensors. Therefore, it becomes crucial to design sensors with high Ingress Protection (IP) ratings like IP65, IP66, or IP67. These sensors need to operate effectively across varying temperature and humidity conditions while maintaining affordability [28]. An overview of some agricultural mobiles with sensors is shown in **Table 1**.

### 3.2 Drought stress detection sensors

RGB sensors can be used to detect plant response to water stress, based on visible parameters [27]. However, many factors can affect RGB imaging, such as lighting, environmental conditions, time of day, and spectral resolution. To overcome these challenges, image processing, and machine learning techniques can be used. Hyperspectral imaging has the potential ability to obtain physiological and phytochemical parameters, which is why it is used to detect drought stress in crops [32]. When these sensors are

| Robot name | Summary of application | Sensors |
|---|---|---|
| Ladybird [29] | Broccoli monitoring | Hyperspectral camera (Resonon Pika XC2), Thermal camera (Xenics Gobi-640-GigE), Stereo camera (GS3-U3–120S6C-C) |
| AgriRover-0 [20] | Maize phenotyping | LiDAR (Velodyne HDL64E-S3) |
| Scout 2.0 [30] | Grapevine phenotyping | LiDAR (RealSense camera VLP16) |
| Robotanist [31] | Sorghum and corn phenotyping | custom stereo camera, point Gray Flea3 cameras |
| RobHortic [12] | Pests and diseases detection in carrot | Multispectral camera (CMS-V), DSLR cameras (EOS 600D), NIR camera (400–1000 nm), Hyperspectral camera (InSpectralVNIR), Thermal camera(A320) |
| XF-ROVIM [13] | *Xylella fastidiosa* detection in olive groves | DSLR cameras (EOS 600D Canon Inc), Multispectral camera (CMS-V Silios), Hyperspectral camera, NIR (spectrograph Inspector V10), Thermal camera (A320), LiDAR (LMS111) |

**Table 1.**
*A summary of agricultural mobile robots and sensors.*

used to detect stress, instead of the spectral information of stressed and non-stressed areas in a complete image of a plant leaf, it separates the affected area and identifies specific imaging patterns and characteristics [33]. Thermal images are the most useful imaging for water stress indicators and drought detection. Still, environmental factors such as solar radiation, air temperature, wind speed, and background temperature can easily affect field measurements [27]. LiDAR acquires various parameters of canopy and leaves, such as vegetation cover, height, canopy structure, leaf area index, and nitrogen status [27]. For using LiDAR sensors in ground-based platforms, integration with GPS and wheel encoder position systems is required for georeferencing. To characterize drought stress in corn, a terrestrial laser scanner has been used to collect LiDAR data, which has the ability to detect three phenotypes related to drought such as plant height, plant area index, and predicted leaf area [34].

## 4. Internet of Robotic Things (IoRT)

IoRT combines the capabilities of robotic devices (such as sensors and actuators) with the Internet of Things (IoT) infrastructure, enabling interconnected and intelligent robotic systems to collect, process, and exchange data autonomously. This convergence allows agricultural robots to operate in dynamic environments, interact with each other and with other IoT-enabled devices, and make data-driven decisions to perform a wide range of tasks across various farming domains. The way robots communicate is influenced by various factors such as the robot, application, and environment which affect the establishment of a communication protocol and the overall performance of the robots. Wireless communication protocols like WiFi [35], ZigBee [36], and Bluetooth [37] are frequently used in digital agriculture. ZigBee is used in agricultural robots that are used in plant health assessment due to hardware availability, low power consumption, and flexible communications models that permit both IP-based and more simplified serial-like messaging. However, they also face drawbacks, including relatively low range and data rates. To address this issue, LoRaWAN emerges as a viable alternative, maintaining the positive attributes of ZigBee for robotic applications while offering extended transmission capabilities with ranges of up to 16 km. Cellular networks like 4G overcome challenges faced by local area network standards by employing a nearly ubiquitous infrastructure-based communication model and benefiting from economies of scale [38]. Now, 4G is available in 85% of global locations [39] and access latency on the order of 50 ms [40]. The 5G network provides a secure communication infrastructure with low latency capabilities. Additionally, with the increasing adoption and success of 3G and 4G, the number of devices supporting 5G is expected to expand as technology advances. This growth is anticipated to lead to the availability of more affordable and energy-efficient devices in the market.

The multi-robot system, a recent technological development, involves the collaboration of mobile robots working together in the agricultural environment. Its purpose is to overcome the limitations associated with tasks that a single mobile robot may be unable to accomplish alone [31]. Multi-robot systems have certain benefits compared to individual mobile robots in precision agriculture applications, including increased effectiveness, efficiency, flexibility, and fault tolerance. Moreover, for efficient operation control, all robots require a reliable and high-throughput network connection with low latency [39]. An example includes the work of [41], in which the development of an autonomous agricultural robot prototype (AgriBot) is intended for various agricultural activities such as seeding, weeding, and spraying fertilizers and insecticides. The robot is equipped with WIFI communication, enabling it to communicate with the operator and other robots. Pretto et al. [42] employed the WiFi protocol for communication between unmanned aerial vehicles and ground vehicles, enabling the monitoring of crop density, weed pressure, and nitrogen (N) nutrition status. The robots were adept at accurately classifying and locating weeds during the spraying process. Additionally, WiFi communication facilitated coordinated missions among the robots, enhancing their collaborative capabilities.

## 5. Machine vision algorithms

In agricultural mobile robots used for plant health assessment, computer vision algorithms, particularly deep learning, play a key role in extracting features from images obtained by sensors. Deep learning algorithms draw inspiration from the functioning of the human brain. Deep learning is a type of neural network trained as a multilayer perceptron (MLP) to learn datasets without manual feature extraction. To generate a deep learning architecture and model, functions are repeated in several layers [43]. These algorithms enable disease detection, weed detection, and yield estimation. An example of such a solution can be found in the work of Blue River Technology. Blue River Technology developed a robotic system called "See & Spray" that uses computer vision and machine learning for plant health assessment. The system employs two sets of camera points, and deep learning algorithms analyze the images to differentiate between crops and weeds. This enables targeted herbicide application, reducing the overall use of herbicides [44]. Ladybird is another example of a robot that uses deep learning for crop detection. The use of Faster R-CNN allows Ladybird to detect and analyze crops efficiently, providing valuable information for monitoring plant health [29]. Weed scouting is an important part of modern integrated weed management. With the advancement of technologies such as machine learning and robots, both time and money can be managed. Hall et al. [45] utilized DCNN data collection algorithms in the AgBotII robot to develop a weed detection model specifically for cotton plants. Similarly, McCool et al. [46] used a lightweight DCNN algorithm to identify weeds. Florance Mary and Yogaraman [47] have applied CNN for detecting weeds between crops and drilling them using a blade attached at the end of the robotic.

Artificial intelligence algorithms play a crucial role in automatic navigation for agricultural robots. A typical agricultural environment has many challenges such as terrain roughness, soil conditions, obstacles, and multiple rows of crops for navigation. This challenge is effectively tackled by training navigation data using machine vision algorithms. In comparison to costly navigation hardware like Real-Time Kinematic Global Positioning System (RTK-GPS), the utilization of machine vision sensors and algorithms offers a cost-effective alternative. This approach enables robots to identify optimal paths and navigate efficiently through agricultural fields [48]. U-Net has been used to process images captured by the Husky robot to predict crop rows and follow the central crop row in a sugar beet field [49]. De Silva et al. [49] presented an agricultural robot based on ANN to make precise path planning. Oliveira et al. [28] have presented their work on the navigation of an agriculture robot in greenhouses. To achieve this, APF, Hector SLAM, and LiDAR are implemented on a robot. AI techniques are capable of minimizing human errors and significantly reducing time consumption by providing data frequently in real-time. In addition, combining various AI techniques can enhance effectiveness in solving agricultural problems.

## 6. Challenges

Robots demonstrate remarkable potential to replace traditional methods for various agricultural tasks and plant care. Despite the advantages of utilizing robots for plant health diagnosis, several challenges remain. For example, identifying pests and weeds necessitates comprehensive data acquisition for each plant, which requires extensive databases to store detailed information. Early detection of plant health issues can avert substantial economic losses in the agricultural sector [50]. Furthermore, vision-based robot and vehicle navigation offer promising prospects for plant health assessment. However, images obtained from agricultural environments such as orchards or farmlands often contain high levels of background noise [51]. Implementing vision navigation systems is challenging due to varying lighting conditions that affect image quality and navigation baseline extraction. Wind can cause plant movement, resulting in blurred images and inaccurate detection [52]. Additionally, terrain slope, soil type, weeds, and dust further complicate vision systems. In addition, most agricultural robots employ 4WD systems, which have drawbacks such as soil compaction and increased dependence on environmental conditions [28]. Additionally, the complex and restrictive regulatory landscape poses obstacles for farmers integrating these technologies. Consequently, further studies are needed to validate the effectiveness of robotics in agriculture. Future research should focus on improving motion systems, computer vision algorithms, and communication technologies for agricultural robots. One recent development for the agricultural environment is the use of legged robots, which can adapt to varying ground slopes and maintain continuous soil contact, thus resisting soil erosion [53]. Additionally, advanced neural networks including convolutional neural networks (CNN) [54] and deep neural networks (DNN) [55] facilitate the identification of diseases and weeds. Deep learning techniques enhance visual information processing, such as image segmentation, feature detection, and target recognition, by reducing environmental noise and vegetation overlap. The integration of IoT technologies with robotic systems can advance smart agriculture by enhancing process control, monitoring, and standardization through artificial intelligence and machine-to-machine communications [56]. This integration holds promise for addressing immediate and long-term agricultural challenges and facilitating the development of precision multi-purpose robotic systems. Furthermore, multi-robot collaboration, involving both ground and aerial vehicles, can enhance vision-based navigation through real-time data exchange and a more comprehensive understanding of the environment.

## 7. Conclusion

This chapter provided insights into enhancing crop management practices through the exploration of agricultural mobile robots equipped with advanced sensing technologies and computer vision algorithms. These robots offer significant advantages over traditional manual methods, including increased efficiency, accuracy, and timeliness in data collection and analysis. However, the agricultural environment's complexity and dynamism present challenges for robots to move beyond small-scale applications or the prototype phase. Research trends indicate that leveraging machine learning techniques enables these robots to effectively analyze data collected on plant health indicators, facilitating early detection and mitigation of plant diseases, nutrient deficiencies, and drought stress. This presents a promising solution for addressing critical challenges in crop production, particularly in plant health assessment and drought stress detection. The adoption of appropriate steering mechanisms tailored to specific agricultural tasks and environmental conditions is crucial for optimizing robot performance and maneuverability. The discussion on mobile robot sensing systems underscores the critical role of sensors in comprehensive data collection and analysis. While technologies such as LiDAR, multispectral imaging, and thermal imaging offer promising capabilities, challenges such as sensor reliability and cost-effectiveness remain to be addressed to ensure the robustness and scalability of sensing systems in agricultural robotics. Additionally, the potential of interconnected robotic systems, facilitated by the Internet of Robotic Things (IoRT), to revolutionize precision agriculture through enhanced communication and collaboration capabilities was highlighted. Despite the transformative potential of agricultural robotics, several challenges must be addressed, including the establishment of comprehensive databases for plant health diagnosis, the development of weather-independent robots, the mitigation of soil compaction, and optimization of AI algorithms for real-time decision-making.

## Acknowledgements

We gratefully acknowledge the funding provided by Adaptive AgroTech Consultancy for this study.

## Author details

Maryam Behjati[1,2], Redmond R. Shamshiri[1,3] and Ibrahim A. Hameed[4*]

1 Adaptive AgroTech Consultancy, Potsdam, Germany

2 Faculty of Agricultural Engineering and Technology, Department of Agricultural Machinery Engineering, University of Tehran, Karaj, Iran

3 Leibniz Institute for Agricultural Engineering and Bioeconomy, Potsdam, Germany

4 Department of ICT and Natural Sciences, Norwegian University of Science and Technology (NTNU), Ålesund, Norway

*Address all correspondence to: ibib@ntnu.no

## References

[1] Lowenberg-DeBoer J, Erickson B. Setting the record straight on precision agriculture adoption. Agronomy Journal. 2019;**111**(4):1552-1569

[2] Ahluwalia O, Singh PC, Bhatia R. A review on drought stress in plants: Implications, mitigation and the role of plant growth promoting rhizobacteria. Resources, Environment and Sustainability. 2021;**5**:100032

[3] Zaidi NW, Dar MH, Singh S, Singh US. Chapter 38—Trichoderma species as abiotic stress relievers in plants. In: Gupta VK, Schmoll M, Herrera-Estrella A, Upadhyay RS, Druzhinina I, Tuohy MG, editors. Biotechnology and Biology of Trichoderma. Elsevier; 2014. pp. 515-525. DOI: 10.1016/B978-0-444-59576-8.00038-2. ISBN 9780444595768

[4] Kaur G, Asthir B. Molecular responses to drought stress in plants. Biologia Plantarum. 2017;**61**(2):201-209. DOI: 10.1007/s10535-016-0700-9

[5] Pautasso M, Dehnen-Schmutz K, Holdenrieder O, Pietravalle S, Salama N, Jeger MJ, et al. Plant health and global change–some implications for landscape management. Biological Reviews. 2010;**85**(4):729-755

[6] Shafi U, Mumtaz R, García-Nieto J, Hassan SA, Zaidi SAR, Iqbal N. Precision agriculture techniques and practices: From considerations to applications. Sensors. 2019;**19**(17):3796

[7] Wang H, Lin Y, Wang Z, Yao Y, Zhang Y, Wu L. Validation of a low-cost 2D laser scanner in development of a more-affordable mobile terrestrial proximal sensing system for 3D plant structure phenotyping in indoor environment. Computers and Electronics in Agriculture. 2017;**140**:180-189

[8] Yang G, Liu J, Zhao C, Li Z, Huang Y, Yu H, et al. Unmanned aerial vehicle remote sensing for field-based crop phenotyping: Current status and perspectives. Frontiers in Plant Science. 2017;**8**:1111

[9] Edan Y, Adamides G, Oberti R. Agriculture Automation. In: Nof SY, editors. Springer Handbook of Automation. Springer Handbooks. Cham: Springer; 2023. DOI: 10.1007/978-3-030-96729-1_49

[10] Atefi A, Ge Y, Pitla S, Schnable J. Robotic technologies for high-throughput plant phenotyping: Contemporary reviews and future perspectives. Frontiers in Plant Science. 2021;**12**:611940

[11] Rubio F, Valero F, Llopis-Albert C. A review of mobile robots: Concepts, methods, theoretical framework, and applications. International Journal of Advanced Robotic Systems. 2019;**16**(2):1729881419839596

[12] Cubero S, Marco-Noales E, Aleixos N, Barbé S, Blasco J. Robhortic: A field robot to detect pests and diseases in horticultural crops by proximal sensing. Agriculture. 2020;**10**(7):276

[13] Rey B, Aleixos N, Cubero S, Blasco J. Xf-Rovim. A field robot to detect olive trees infected by Xylella fastidiosa using proximal sensing. Remote Sensing. 2019;**11**(3):221

[14] Xu R, Li C. A review of high-throughput field phenotyping systems: Focusing on ground robots. Plant Phenomics. 2022;**2022**. DOI:10.34133/2022/9760269

[15] Iqbal J, Xu R, Halloran H, Li C. Development of a multi-purpose autonomous differential drive mobile robot for plant phenotyping and soil sensing. Electronics. 2020;**9**(9):1550

[16] Mueller-Sim T, Jenkins M, Abel J, Kantor G. The Robotanist: A ground-based agricultural robot for high-throughput crop phenotyping. In: 2017 IEEE International Conference on Robotics and Automation (ICRA). Singapore; 2017. pp. 3634-3639. DOI: 10.1109/ICRA.2017.7989418

[17] Shafiekhani A, Kadam S, Fritschi FB, DeSouza GN. Vinobot and vinoculer: Two robotic platforms for high-throughput field phenotyping. Sensors. 2017;**17**(1):214

[18] Zhang H, Zhang Y, Yang T. A survey of energy-efficient motion planning for wheeled mobile robots. Industrial Robot: The International Journal of Robotics Research and Application. 2020;**47**(4):607-621

[19] Nguyen P, Badenhorst PE, Shi F, Spangenberg GC, Smith KF, Daetwyler HD. Design of an unmanned ground vehicle and lidar pipeline for the high-throughput phenotyping of biomass in perennial ryegrass. Remote Sensing. 2020;**13**(1):20

[20] Qiu Q, Sun N, Bai H, Wang N, Fan Z, Wang Y, et al. Field-based high-throughput phenotyping for maize plant using 3D LiDAR point cloud generated with a "phenomobile". Frontiers in Plant Science. 2019a;**10**:554. DOI: 10.3389/fpls.2019.00554

[21] Botta A, Cavallone P, Baglieri L, Colucci G, Tagliavini L, Quaglia G. In depth analysis of power balance, handling, and the traction subsystem of an articulated skid-steering robot for sustainable agricultural monitoring. SN Applied Sciences. 2023;**5**(4):103

[22] Colucci G, Botta A, Tagliavini L, Cavallone P, Baglieri L, Quaglia G. Kinematic modeling and motion planning of the mobile manipulator Agri.Q for precision agriculture. Machines. 2022;**10**(5):321

[23] Gai J, Xiang L, Tang L. Using a depth camera for crop row detection and mapping for under-canopy navigation of agricultural robotic vehicle. Computers and Electronics in Agriculture. 2021;**188**:106301

[24] Sori H, Inoue H, Hatta H, Ando Y. Effect for a paddy weeding robot in wet rice culture. Journal of Robotics and Mechatronics. 2018;**30**(2):198-205

[25] Biber P, et al. Navigation System of the Autonomous Agricultural Robot Bonirob. Workshop on Agricultural Robotics: Enabling Safe, Efficient, and Affordable Robots for Food Production (Collocated with IROS 2012), Vilamoura, Portugal. 2012

[26] Young SN, Kayacan E, Peschel JM. Design and field evaluation of a ground robot for high-throughput phenotyping of energy sorghum. Precision Agriculture. 2019;**20**(4):697-722

[27] Kim J, Kim K, Kim Y-H, Chung YS. A short review: Comparisons of high-throughput phenotyping methods for detecting drought tolerance. Scientia Agricola. 2020;**78**(4)1-8. DOI: 10.1590/1678-992x-2019-0300

[28] Oliveira LF, Moreira AP, Silva MF. Advances in agriculture robotics: A state-of-the-art review and challenges ahead. Robotics. 2021;**10**(2):52. DOI: 10.3390/robotics10020052

[29] Bender A, Whelan B, Sukkarieh S. A high-resolution, multimodal data set for agricultural robotics: A Ladybird's-eye view of Brassica. Journal of Field Robotics. 2020;**37**(1):73-96. DOI: 10.1002/rob.21877

[30] Tiozzo Fasiolo D, Pichierri A, Sivilotti P, Scalera L. An analysis of the effects of water regime on grapevine canopy status using a UAV and a mobile

robot. Smart Agricultural Technology. 2023;**6**:100344. DOI: 10.1016/j.atech.2023.100344

[31] Zhang C, Noguchi N, Yang L. Leader–follower system using two robot tractors to improve work efficiency. Computers and Electronics in Agriculture. 2016;**121**:269-281. DOI: 10.1016/j.compag.2015.12.015

[32] Behmann J, Steinrücken J, Plümer L. Detection of early plant stress responses in hyperspectral images. ISPRS Journal of Photogrammetry and Remote Sensing. 2014;**93**:98-111. DOI: 10.1016/j.isprsjprs.2014.03.016

[33] Zubler AV, Yoon J-Y. Proximal methods for plant stress detection using optical sensors and machine learning. Biosensors. 2020;**10**(12):193

[34] Su Y, Wu F, Ao Z, Jin S, Qin F, Liu B, et al. Evaluating maize phenotype dynamics under drought stress using terrestrial lidar. Plant Methods. 2019;**15**:11. DOI: 10.1186/s13007-019-0396-x

[35] Hajjaj SSH, Sahari KSM. Review of agriculture robotics: Practicality and feasibility. In: 2016 IEEE International Symposium on Robotics and Intelligent Sensors (IRIS). Tokyo, Japan; 2016. pp. 194-198. DOI: 10.1109/IRIS.2016.8066090

[36] Bodunde O, Adie U, Ikumapayi O, Akinyoola J, Aderoba A. Architectural design and performance evaluation of a ZigBee technology based adaptive sprinkler irrigation robot. Computers and Electronics in Agriculture. 2019;**160**:168-178

[37] Sowjanya KD, Sindhu R, Parijatham M, Srikanth K, Bhargav P. Multipurpose autonomous agricultural robot. In: 2017 International Conference of Electronics, Communication and Aerospace Technology (ICECA). Coimbatore, India; 2017. pp. 696-699. DOI: 10.1109/ICECA.2017.8212756

[38] Gielis J, Shankar A, Prorok A. A critical review of communications in multi-robot systems. Current Robotics Reports. 2022;**3**(4):213-225. DOI: 10.1007/s43154-022-00090-9

[39] Zhivkov T, Sklar EI, Botting D, Pearson S. 5G on the farm: Evaluating wireless network capabilities and needs for agricultural robotics. Machines. 2023;**11**(12):1064. Available from: https://www.mdpi.com/2075-1702/11/12/1064

[40] Tahir MN, Katz M. Performance evaluation of IEEE 802.11p, LTE and 5G in connected vehicles for cooperative awareness. Engineering Reports. 2022;**4**(4):e12467. DOI: 10.1002/eng2.12467

[41] Amer G, Mudassir SMM, Malik MA. Design and operation of Wi-Fi agribot integrated system. In: 2015 International Conference on Industrial Instrumentation and Control (ICIC); Pune, India; 28-30 May 2015. 2015. pp. 207-212. DOI: 10.1109/IIC.2015.7150739

[42] Pretto A, Aravecchia S, Burgard W, Chebrolu N, Dornhege C, Falck T, et al. Building an aerial-ground robotics system for precision farming: An adaptable solution. arXiv e-prints, arXiv:1911.03098. 2019. DOI: 10.48550/arXiv.1911.03098

[43] Alzubaidi L, Zhang J, Humaidi AJ, Al-Dujaili A, Duan Y, Al-Shamma O, et al. Review of deep learning: Concepts, CNN architectures, challenges, applications, future directions. Journal of Big Data. 2021;**8**(1):53. DOI: 10.1186/s40537-021-00444-8

[44] Yeshe A, Gourkhede P, Vaidya PH. Blue River Technology: Futuristic

Approach of Precision Farming. Punjab, India: Just Agriculture; 2022

[45] Hall D, Dayoub F, Kulk J, McCool C. Towards unsupervised weed scouting for agricultural robotics. In: Chen IM, Nakamura Y, editors. Proceedings of the 2017 IEEE International Conference on Robotics and Automation (ICRA). Singapore; 2017. pp. 5223-5230. DOI: 10.1109/icra.2017.7989612

[46] McCool C, Perez T, Upcroft B. Mixtures of lightweight deep convolutional neural networks: Applied to agricultural robotics. IEEE Robotics and Automation Letters. 2017;**2**(3):1344-1351. DOI: 10.1109/LRA.2017.2667039

[47] Florance Mary M, Yogaraman D. Neural network based weeding robot for crop and weed discrimination. Journal of Physics: Conference Series. 2021;**1979**(1):012027. DOI: 10.1088/1742-6596/1979/1/012027

[48] Shamshiri R, Hameed I, Pitonakova L, Weltzien C, Balasundram S, Yule I, et al. Simulation software and virtual environments for acceleration of agricultural robotics: Features highlights and performance comparison. International Journal of Agricultural and Biological Engineering. 2018;**11**:15-31. DOI: 10.25165/ijabe.v11i4.4032

[49] De Silva R, Cielniak G, Wang G, Gao J. Deep learning-based crop row detection for infield navigation of agri-robots. Journal of Field Robotics. 2023;**32**:162-175

[50] Chapman SC, Merz T, Chan A, Jackway P, Hrabar S, Dreccer MF, et al. Pheno-copter: A low-altitude, autonomous remote-sensing robotic helicopter for high-throughput field-based phenotyping. Agronomy. 2014;**4**(2):279-301. Available from: https://www.mdpi.com/2073-4395/4/2/279

[51] Chen M, Tang Y, Zou X, Huang Z, Zhou H, Chen S. 3D global mapping of large-scale unstructured orchard integrating eye-in-hand stereo vision and SLAM. Computers and Electronics in Agriculture. 2021;**187**:106237

[52] Mavridou E, Vrochidou E, Papakostas GA, Pachidis T, Kaburlasos VG. Machine vision systems in precision agriculture for crop farming. Journal of Imaging. 2019;**5**(12):89

[53] Oliveira LFP, Rossini FL. Modeling, simulation and analysis of locomotion patterns for hexapod robots. IEEE Latin America Transactions. 2018;**16**:375-383

[54] Emmi L, Le Fl'echer E, Cadenat V, Devy M. A hybrid representation of the environment to improve autonomous navigation of mobile robots in agriculture. Precision Agriculture. 2021;**22**(2):524-549

[55] Adhikari SP, Kim G, Kim H. Deep neural network-based system for autonomous navigation in paddy field. IEEE Access. 2020;**8**:71272-71278

[56] Cui F. Deployment and integration of smart sensors with IoT devices detecting fire disasters in huge forest environment. Computer Communications. 2020;**150**:818-827

Chapter 2

# Use Cases of Technologies in Precision Agriculture: Selected Abstracts Submitted to the 10th Asian-Australasian Conference on Precision Agriculture (ACPA10)

*Redmond R. Shamshiri, Maryam Behjati, Siva K. Balasundram, Christopher Teh Boon Sung, Ibrahim A. Hameed, Ahmad Kamil Zolkafli, An Ho-Song, Arina Mohd Noh, Badril Hisham Abu Bakar, W.A. Balogun, Beom-Sun Kang, Cong-Chuan Pham, Dang Khanh Linh Le, Dong Hee Noh, Dongseok Kim, Eliezel Habineza, Farizal Kamaroddin, Gookhwan Kim, Heetae Kim, Hyunjung Hwang, Jaesung Park, Jisu Song, Joonjea Sung, Jusnaini Muslimin, Ka Young Lee, Kayoung Lee, Keong Do Lee, Keshinro Kazeem Kolawole, Kyeong Il Park, Longsheng Fu, Md Ashrafuzzaman Gulandaz, Md Asrakul Haque, Md Nasim Reza, Md Razob Ali, Md Rejaul Karim, Md Sazzadul Kabir, Md Shaha Nur Kabir, Minho Song, Mohamad Shukri Zainal Abidin, Mohammad Ali, Mohd Aufa Md Bookeri, Mohd Nadzim Nordin, Mohd Nadzri Md Reba, Mohd Nizam Zubir, Mohd Saiful Azimi Mahmud, Mohd Taufik Ahmad, Muhammad Hariz Musa, Muhammad Sharul Azwan Ramli, Musa Mohd Mokji, Naoto Yoshimoto, Nhu Tuong An Nguyen, Nur Khalidah Zakaria, Prince Kumar, P.K. Garg, Ramlan Ismail, Ren Kondo, Ryuta Kojo, Samsuzzaman, Seokcheol Yu, Seok-Ho Park, Shahriar Ahmed,*

*Siti Noor Aliah Baharom, Sumaiya Islam, Sun-Ok Chung, Ten Sen Teik, Tinah Manduna Mutabazi, Wei-Chih Lin, Yeon Jin Cho and Young Ho Kang*


## Abstract

This chapter is a collection of selected abstracts presented at the 10th Asian-Australasian Conference on Precision Agriculture, held from October 24th to 26th in Putrajaya, Malaysia. It aims to emphasize the transformative potential of technology in precision agriculture and smart farming. The featured studies highlight the transformative impact of technology and current improvements in agriculture, offering modern solutions including machine learning, robotics, remote sensing, and geographic information systems (GIS). From autonomous navigation for mobile robots to stress classification in crop production systems, and from phenotypic analysis with LiDAR technology to real-time sensor monitoring in greenhouse agriculture, the majority of abstracts underline the integration of digital tools in different fields of farming with the core objective of reshaping conventional farming techniques and eliminating dependency on manual works. Key examples include the development of a distributed sensing system (DSS) used for orchard robots, stress classification for tomato seedlings through image-based color features and machine learning, and the integration of remote sensing and AI in crop protection. Other solutions, such as automated spraying robots for cherry tomato greenhouses, active back exoskeletons for rice farm lifting tasks, and advancements in seedling transplanting techniques, have shown promising results for contributing to sustainable farming practices by providing accurate and timely information for decision-making amid climate change-induced uncertainties.


**Keywords:** digitalization, IoT, artificial intelligence, smart sensors, robotics, yield optimization

## 1. Introduction

Precision agriculture has witnessed a significant transformation driven by advancements in technology and data-driven approaches such as machine learning, robotics, remote sensing, and geographic information systems (GIS), emerging as a basis for revolutionizing traditional farming methods. This pattern shift is evident in various aspects of farming practices, including navigation systems and path planning algorithms for efficient operation of mobile robotics, stress classification in open field and closed field crop production systems such as plant factories, phenotypic analysis using LiDAR technology, and enhancement of crop yields through remote sensing and artificial intelligence. In addition, innovations such as algae cultivation monitoring, volumetric yield estimation using RGB cameras, intelligent decision support systems, and real-time sensor for early monitoring of greenhouse diseases are reshaping data collection and decision makings in modern agriculture.

Drawing from a diverse array of abstracts submitted to the 10th Asian-Australasian Conference on Precision Agriculture, this chapter presents 26 selected studies showcasing the transformative potential of technology in modern agriculture and highlighting cutting-edge advancements. The main critical challenges addressed by these advancements are the need for precision and efficiency in different farming operations. Examples such as a distributed sensing system (DSS) that aids in the precise navigation of an orchard robot through the integration of GPS, laser, and infrared sensors; methods for improving route optimization and mitigating crop damage; and an obstacle region segmentation approach that streamlines path planning in agricultural machinery operations, all serve to minimize redundancy and enhance efficiency. For crop protection, stress classification in tomato seedlings has been tested through image-based color features and machine learning, facilitating early stress detection for optimal plant growth. Additionally, the utilization of 3D LiDAR technology in fruit tree phenotyping has accelerated and refined orchard characteristic assessment, thereby assisting in informed decision-making. The integration of remote sensing, GIS, and AI in crop protection offers data-driven insights for mapping crop types, health assessment, and yield estimation, thereby revolutionizing risk management strategies in agriculture.

Traditional methods often fail to provide accurate and live information required for rapid decision-making, especially in the face of climate change-induced uncertainties and environmental risks. However, the integration of breakthrough technologies such as drones, multispectral imaging, machine learning, and sensor networks is revolutionizing how farmers monitor, analyze, and manage their crops. Moreover, the application of data fusion techniques and predictive modeling is empowering farmers with real-time insights and cloud-based decision support tools. By integrating data from diverse sources such as weather forecasts, satellite imagery, and soil sensors, intelligent systems can provide personalized recommendations for irrigation, fertilization, pest control, and harvest timing. These data-driven approaches not only optimize resource utilization but also enhance crop yields and reduce production costs. Through the utilization of environmental sensors and real-time data analysis, researchers are developing disaster-resistant structures for greenhouses, mitigating the impact of weather disasters like droughts, floods, and heat waves. These solutions facilitate continuous monitoring of crop conditions, ensuring stable and high-quality production even in adverse weather conditions.

The abstracts featured in this chapter demonstrate how innovative solutions and emerging technologies are enhancing agricultural efficiency, resilience to external factors, and reducing reliance on manual labor, while maximizing profitability.

## 2. A distributed sensing system for assisted navigation of mobile robots

Autonomous navigation of agricultural mobile robots through uneven terrains, avoiding obstacles, and optimizing routes are intricate challenges that demand innovative solutions. Conventional navigation methods, often reliant on GPS alone, face limitations in accuracy and reliability, especially in densely vegetated areas with poor satellite reception and disturbances of the outdoor environment. Data fusion and multiple perception solutions are usually employed to assist the existing RTK GPS-based navigation and to improve the reliability of the operation in unstructured agricultural fields. In this regard, a distributed sensing system (DSS) that leverages multiple data sources can mitigate these limitations and maintain reliable navigation even in challenging conditions. This study reports on the design, development, and evaluation of a DSS with CANBUS communication to assist the autonomous navigation of a four-wheel steering agricultural mobile robot inside berry orchards. The proposed DSS was responsible for detecting random obstacles and providing the navigation controller with real-time feedback for effective collision avoidance. For the proof of concept, the assisted navigation was expected to maintain the robot between the plants' rows with an accuracy of 10–20 cm from the side with an ideal forward speed of 5–8 km/h. In the hardware layer, the DSS benefits from a Jetson Nano onboard computer along with a set of ROS-based multi-channel infrared and laser sensors for perception. In the software layer, a fuzzy knowledge-based algorithm was designed and simulated. Results demonstrated significant improvements in navigation accuracy through the fusion of GPS, laser, and infrared sensor data, which was essential to avoid crop damage and ensure optimal resource utilization. Results from simulation and field experiments also suggested that the exponential filter was necessary to be implemented on each sensor for removing noise and outliers. It was concluded that the development of such systems requires extensive validation tests in different orchards besides a more accurate dynamic model of the robot platform. This can be accelerated using digital models inside virtual replicas of the environments.

The base vehicle was a combustion engine four-wheel drive (4WD) and four-wheel steering (4WS) Quatrak manufactured by Irus (IRUS Motorgeräte GmbH, Burladingen, Germany) with a track width of 1.2 m, maximum forward and reverse speed of 10 km/h, and approximate weight of 475 kg that utilizes a 2-cylinder petrol engine with 20.1 kW (27 HP) output power. The 4WS mechanism of the vehicle allowed a minimum turning radius on the row-ends with increased precision and control over the robot's movement, which is of great importance for making sharp turns and navigating in tight spaces inside orchards. The vehicle benefits from the HYDAC TTC 580 (HYDAC, Sulzbach, Germany) controller to translate and deliver the control steering signals to the actuators (including the electrohydraulic system) that are responsible for speed and turning angle. A GPS-based navigation toolbox was installed on the vehicle by Innok Robotic (Innok Robotics GmbH, Regenstauf, Germany) that benefits from a custom-built graphical user interface for uploading waypoints and following trajectory. In order to validate the proposed assisted navigation system, field visits were first carried out to collect preliminary data using high-accuracy RTK GPS. These data were used to create a virtual orchard inside CoppeliaSim that was interfaced with ROS for testing different sensors, hardware in the loop, and control algorithms on a full-scale simulated robot and orchard model. The main elements of the simulation scenes in this project were (i) mesh files representing plants, the robot, and obstacles, (ii) API and codes that created interfaces between different software environments, and (iii) algorithms and dynamic models including minimum distance calculation, Ackerman steering, path following, and obstacle avoidance algorithms. The simulation approach converted native data streams from various sensor inputs into usable information within the command-and-control system. The sensors used in front of the robot were TFmini Plus single-point short-range LiDAR sensor manufactured by Benewake (Benewake, China) with a distance detection range between 0.02 and 12 m, resolution of 1 cm, sampling rate of 100 Hz, frame rate between 1 and 1000 Hz, accuracy of $\pm 1\%$, and field of view of 3.6°. In the communication layer, CANBUS (ISO 11898-2) was used for exchanging data between different nodes due to its scalability and high reliability for automation even in harsh environments with high electromagnetic interference.

The simulation scene provided a safe, fast, and low-cost experiment platform for the development, testing, and validating of the sensing and control strategies with different algorithms. It also enabled human-aware navigation by finding the best positions for each sensor and provided a flexible solution for attaching other implements and determining the optimum row-end turning patterns in the presence of random obstacles. It also accelerated complicated analysis to understand the behavior of the robot on uneven terrains. Results of field experiments showed the reliable functionality of these sensor arrays under outdoor light conditions to detect the bushes and random obstacles within the pre-programmed range of 5–60 cm. In addition to the detection task, each sensor module also transmits a state signal equal to 0 or 1 on the CANBUS line in the absence or presence of an obstacle that is georeferenced with the robot position for monitoring purposes. The average distance between the robot and the left bushes, as measured by the three sets of IR sensors mounted on the left of the robot were 17.5, 18.5, and 17.1 cm. These values for the right bushes were found to be 13.5, 18.9, and 16.6 cm. It was observed that the front IR sensor provided the collision avoidance controller with accurate feedback to stop the robot from nearly a 25 cm distance from the bushes. These results show the redundancy in the distance measurements that the multi-channel IR sensing setup are providing to make sure the controller always receives feedback and the robot continues its operation, even if one sensor module fails. Field experiments also confirmed that the analog output of these sensors with the exponential filter noise reduction allows for precise measurements, even in high-density bush conditions, and the multi-channel ensured that the final measurements were not easily affected by the disturbances of the environment such as occlusion by leaves and dust. From the perspective of exchanging sensor data with the controller, the CANBUS network was found to be robust and reliable with 100% of the data transmitted during the experiment without any interruption.

The system showcased consistent performance across different types of crops, soil conditions, and environmental factors. The hardware of the system was designed to be modular and could be replaced easily upon failure. This adaptability stems from the fusion of multi-modal sensor data, which allows the robot to gather comprehensive information and make informed decisions regardless of the surroundings. Comparative analysis with existing navigation methods highlighted the superiority of the DSS. The quantitative comparisons demonstrated the system's significant advantages in terms of accuracy, efficiency, and adaptability. The system's ability to generalize its navigation strategies to various contexts holds great promise for its widespread adoption in agriculture. The evaluation confirmed that the proposed solution is capable of adjusting the robot's steering and speed steering the robot in the presence of different obstacles and in different scenarios. This was achieved by a control strategy that reduced the error distance between the sensor readings and a minimum allowed distance with left, right, and front obstacles.

## 3. Obstacle-driven path planning via region segmentation

The efficiency of agricultural machinery operation is improved to satisfy the needs of autonomous agricultural machinery operation in the field. Aimed at the challenges of farming operations such as obstacles, overlapping and redundant working paths, and a higher number of turns. This paper proposes a complete coverage path planning algorithm based on obstacle region segmentation. For obstacles in the environment map, the minimum rectangle algorithm is used to segment the obstacles and achieve the segmentation of obstacle areas and obstacle-free areas. The polygon segmentation algorithm was used to divide the obstacle-free areas into working sub-areas. The reciprocating algorithm was applied to cover the sub-areas and generate the coverage path. The Dijkstra algorithm was utilized to establish the connecting paths between adjacent sub-areas to achieve complete coverage path planning of the entire farmland. Several experiments were conducted to verify the feasibility and practicality of the proposed approach, which successfully avoids obstacles, reduces path redundancy, and improves coverage.

## 4. Tomato seedling stress classification in plant factory with machine learning

Controlling key environmental parameters, such as temperature, relative humidity, light intensity, and water quantity, is crucial for ensuring optimal plant growth. In a controlled environment, early detection and management of environmental stress play a vital role in achieving high-yield and high-quality tomato seedlings. However, conventional invasive techniques for detecting plant stress are impractical and susceptible to inaccuracies. This study aimed to employ image-based color feature characteristics and machine learning models for accurate classification of stress symptoms in tomato seedlings. Two-week-old tomato seedlings were placed into five chambers in a controlled plant factory, and the seedlings were grown at different temperatures (20, 25, and 30°C), light intensity (50, 250, 450 PPFD), day and night hours (8/16, 10/14, and 16/8), and water supply levels (1, 2, and 3 L/day). Humidity (60%), $CO_2$ level (600–800 ppm), and nutrient treatment (0.8 dS/m) were constant for all the chambers. A microcontroller was utilized for automated image acquisition from each chamber and store images of tomato seedlings. Nine textures and 18 color features were extracted for each image and a two-way analysis of variance was used to determine the statistical significance of treatment effects. Duncan's multiple range test, and sequential forward floating selection (SFFS) algorithms were used to select features and reduce feature overlaps. Support vector machine (SVM), naive Bayes classifier (NBC), K-nearest neighbor (KNN), and random forest classifier (RFC) were employed to identify the optimal classifier for stress and non-stress classification of tomato plants. The results indicated the impact of stress on seedlings became noticeable approximately 3 days after seedling placement. Morphological analysis demonstrated significant differences in stress conditions attributed to temperature and water content variations, whereas light intensity and day-night hours exhibited less significant differences (<5%) on Duncan's multiple range test. The SVM, KNN, NBC, and RFC achieved average stress classification accuracies of 89.56%, 89.23%, 79.05%, and 63.75%, respectively. The combination of image-extracted features and machine learning models can improve environmental stress monitoring in plant factories, leading to better crop quality, yield, and resource utilization. Future research should explore additional factors and expand the dataset for further improvements.

## 5. Orchard fruit tree phenotyping with 3D LiDAR quantification

Phenotypic information plays a crucial role in effective fruit tree management in orchards. The utilization of LiDAR (Light detection and ranging)-based object recognition provides a rapid and precise assessment of the phenotypic traits of orchard fruit trees. This study aimed to quantify the fruit trees phenotypic and land characteristics (e.g., tree height, canopy area, tree spacing, and inter-row spacing) in an apple orchard using a 3D LiDAR. A LiDAR sensor (i.e., VLP-16) was employed to collect 3D point cloud data from an apple orchard. Six apple trees, exhibiting diverse shapes and sizes, were selected for testing purposes. Commercial software was utilized to process the collected data for visualization and measurements. The accuracy of the estimated outputs from the point cloud was evaluated by comparison with measured values. The estimated tree heights from the point cloud were 3.05 ± 0.34 m, while the measured heights were 3.13 ± 0.33 m. The root mean square error (RMSE) for tree height estimation was 0.09 m, and the simple linear coefficient of determination ($r^2$) was 0.98. Regarding the canopy area, point cloud estimations were 5.97 ± 1.4 $m^2$, compared to the measured area of 6.19 ± 1.38 $m^2$. The RMSE for canopy area estimation was 0.24 $m^2$, with $r^2$ of 0.99, indicating a high level of accuracy. For tree spacing and row distance, the point cloud measurements were 3.04 ± 0.17 m, and 3.18 ± 0.24 m, respectively, while the measurements were 3.35 ± 0.08 m, and 3.40 ± 0.05 m, respectively. The RMSE and $r^2$ for tree spacing were 0.12 m and 0.92, respectively, and for row distance, 0.07 m and 0.94, respectively. Despite minor differences, there was a strong relationship and close agreement between the estimation using point cloud data and measurements. The findings highlight the reliability and efficiency of the 3D LiDAR technology for accurately detecting and measuring phenotypic traits in fruit trees. This research would help in quantification of fruit trees phenotypic and land characteristics for maximizing fruit cultivation.

## 6. Crop insurance enhancement with RS, GIS, GEE, and AI

This paper discusses remote sensing applications, the Google Earth Engine (GEE), and the use of artificial intelligence (AI) in crop insurance. The global crop insurance market is vast and diverse. However, remote sensing applications for crop insurance are already in use. There is a possibility of combining remote sensing, GEE, and AI by insurance companies considering the insured amount of the crop. This study explores the capability of remote sensing for payout for the insured crops. Geospatial tools are capable of mapping and giving the basic requirements of the crop insurance assistance system. The process of the crop insurance assistance system requires the agro-climate, crop calendar, crop yield predictions, a record of crop rotations, soil productivity, crop classification, crop monitoring, types of damage to the crop, crop damage assessment, and finally the crop insurance settlement amount. In the past, traditional methods of ground surveying could be relied upon up to a certain point to obtain this type of information. Remote sensing combined with the GEE and AI provides a more dependable and efficient method of gathering the necessary data for mapping crop types and estimating the acreage, and crop loss. This information can be very helpful for agricultural-based insurance/finance companies as well as farmers. By utilizing multispectral sensors and GEE, it is possible to analyze the spectral reflectance of crops over time and derive meaningful information about their health and vigor as well as damage. Different spectral bands of the sensors provide valuable data for calculating various vegetation indices, such as the normalized difference vegetation index (NDVI) or the enhanced vegetation index (EVI). The spectral reflectance curves may vary during the growth and health conditions of crops, and thus measurement and monitoring can be done by multispectral sensors. Radar data are sensitive to the structure of crops, crop moisture content, and alignment. It would provide information regardless of the time or weather, and thus offer additional information to optical sensors. Combining both the sensors' data would expand the available information for making a distinction between each crop class and its spectral signature. As a result, there seems to be a greater chance of achieving higher classification accuracy of yield, and crop damage assessment required for payout for crop insurance.

## 7. Spatial-temporal analysis and prediction in Ugandan coffee farming

Uganda contributes 3% of global production of both the Arabica and Robusta coffee varieties, uncontested ranking second in Africa for coffee export. Due to climate change coffee is highly vulnerable. Rising temperatures and increasingly erratic rainfall have exposed trees to pests and diseases, decreasing both the quantity and quality of the crop. Uganda Coffee Development Authority (UCDA) lacks a reliable and fast mechanism for monitoring coffee crops to quickly identify crop challenges and hence has a low chance of combating the risks of depletion of potential yields in a timely manner. Drones have proved to be precise as compared to the conventional method of assessments that are time-consuming and impossible to examine all the crops deriving to area sampling. Therefore, Uganda Flying Labs (UFL) collaborated with UCDA to provide support turnaround solutions. Multispectral and RGB drone flights were significant in capturing UAV images in support of coffee crop health problems ranging from plants under stress, due to a water or fertilizer shortage, or pest attacks, inexorably saving coffee plants from premature death and ensuring high coffee production. The study enabled yield projections, damage assessment, flood impact, and NDVI through the use of UAV images for data analysis. Successive progress has been made from UAV coffee mapping results intersecting with terrain formation data affecting crop growth. Use of GIS tools such as deep learning in ArcGIS Pro managed to track crop health, inventory, and preharvest projections. The study gave an experience in using the RCNN mask deep learning technique in modeling for object identification of intercropped coffee with banana plantations. The artificial intelligence application assisted in crop inventory ultimately giving a snippet of coffee beans anticipated from the mature crops. Spatial and temporal variability and prediction analysis aspects to mitigate the gaps caused by erosion and unfavorable rain patterns influenced effective response. Crop health, statistics, and harvest estimates were the standard in the whole process.

A coffee farm (Ventura) of 44.034 ha (0.44 $km^2$) was identified. Its location was on a hillside in a moderately hot central region, namely Mukono. The coffee plantation holds different ages of robust species intercropped with banana plants in some acreage to give shade. The data input was achieved through drone flights of a Phantom 4 for RGB images and a DJI Phantom 4 Multispectral with six sensors 1/2.9″ CMOS, including one RGB sensor for visible light imaging and five monochrome sensors for multispectral imaging. An RGB-dedicated drone was important to use because of its high resolution compared to multispectral to facilitate accurate predictions and modeling through algorithms. The flight plans were PIX4D Capture and DJI GS Pro for Phantom 4 and Multispectral respectively at flight height of 75 m with a GSD of 4 cm/pix ultimately acquiring 10,722 images. The task took 2 days to be accomplished with four flights from the multispectral while onboarding RTK enabled GPS for georeferencing. The RGB drone completed its work with two flights. Data processing was carried out with Pix4Dmapper and PIX4D fields to develop Orthomosaic, DSM, and reflectance images for individual bands. Furthermore, analysis was executed with ArcGIS Pro and PIX4D Survey to produce Coffee plants (shapefile) with attributes of age and health status, DTM, and slope map including NDVI map. The final output shared with the stakeholders was a map in PDF format of the Orthophoto, Terrain (contours), Slope, NDVI, Crop Health & Harvest Estimates. The coffee farm was allotted according to coffee age and crops were automatically detected using the RCNN mask deep learning technique in ArcGIS Pro for health status ranging from very poor to good crop. Estimates were in density per hectare and harvest tonnage in kilograms.

UCDA was keen on a comparison of both methods of cropping which is intercropped coffee with a shading of banana plants and the coffee trees that were planted independently. This would help them to use the results to decide on best practices for different terrains. With the time not on our side, we were given the option of a commercial farmer who had both types of farming methods on a rocky and hilly piece of land. Reporting figures of yield steered interest in analyzing how technology could help in the estimation of yields nationwide and efficiently address factors limiting improvements in those yields, subsequently for exports. While the bank looked forward to the potential to strengthen relationships with their clients by facilitating informed decisions making with timely and targeted interventions to maximize return on investments (loans given to farmers). The generated digital surface model, reflectance maps for the various bands, and a 3-D Point Cloud with further processing, more results such as the Digital Terrain Model slope map, NDVI maps an indicator of crop health presented a focus to the farmer, UCDA of best practices. Highly acclaimed deep learning techniques for automatic object identification in an image and sampling were evident. With a few iterations to the model, the model was able to detect over 7000 stands. The tool proved useful for inventorying and helped in coming up with further analysis focused on the identified stands. Identified stands, overlayed with the NDVI map, the status of each crop was assessed, which enabled the indication of areas with stress. Therefore, a known number of crop stands and harvest estimates were calculated based on the expected yield per year of mature coffee trees. An estimate of the harvest was 8500 kg of green beans (dried coffee beans without husks) in the range of approximately 7.5 Ha was the generated results. The use of a Digital Terrain Model, assessment of erosion and flood risk data, and contour maps empowered the farmer to make a decision on drip line construction and terracing as some of the solutions.

The outcomes of this study revealed that spatial characteristics are relational to weather patterns consequently affecting crop health. Drone technology and data complimented by RCNN Masks deep learning model automatically detected coffee crops proving to be a quick solution to mitigate risks in coffee farming. Harvest prediction analysis presented through NDVI, slope maps, and crop health index, reflected the volume of yields that were expected conclusively reflecting a number of trees to be replaced. The risk of erosion enabled the farmer to mitigate through soil retaining techniques on slopes to prevent the lose of soil and nutrients. With the evidence of spatial and temporal variability, modeling, and prediction analysis in coffee farming there is room for UCDA to scale up drone data applications to improve coffee production in Uganda. It can be concluded that: "aerial platforms data provides an efficient, rapid and nondestructive assessment through the spectral response of plants and its relationship with various environmental factors [1]."

## 8. Algae cultivation and harvesting

This work describes a novel approach for cultivating and assessing algae growth in various aquatic systems. The study utilizes an unmanned aerial vehicle equipped with a multispectral camera to detect and monitor the presence and growth of the algal community. The aim is to provide an efficient and non-invasive solution for evaluating algae growth using vegetative and other indices. The research also focuses on developing a predictive model for algae growth through correlation analysis, which is particularly effective in image analysis tasks. The model demonstrates high precision in measuring and predicting algal growth stages across different environments, compensating for limitations associated with traditional data collection methods. The findings of this study have the potential to significantly impact algae growth measurement and prediction, contributing to the advancement of sustainable and economically viable practices. The results will be compared with alternative measurement methods for further evaluation.

## 9. Volumetric yield estimation of Chinese cabbage using RGB camera

Yield assessment helps farmers to make efficient use of their resources for an accurate estimation of crop production. The objective of this research was to determine the volume of Chinese cabbage using a low-resolution RGB camera on an application board (i.e., Raspberry Pi). The system consists of two 9-W fluorescent LED light sources, two ultrasonic sensors, three low-resolution cameras, and an HP Core i7 laptop to capture RGB images of 30 cabbage samples. The camera and LED lighting were mounted 1.08 meters above the harvester conveyor belt. The speeds of the conveyor were 0.55, 0.70, and 0.85 m/s, respectively. Two ultrasonic sensors triggered the camera at a rate of 5 Hz to capture RGB images of the samples. The actual volumes of the 30 cabbages were calculated using the standard method of Archimedes law, and a combination of RGB image processing and volume estimation techniques for cabbage were developed and compared. The volume of cabbage with an ellipsoidal shape was estimated using the box method. In this method, a Python-based program was developed to calculate the area of the RGB images by background subtraction, edge detection, area calculation, and pixel value count, and the height of each cabbage was determined by subtracting each cabbage sample point from the background point. The linear regression method and t-test analyses at a 5% level were conducted to compare the estimated and measured cabbage volumes. These conventional and box estimation approaches provided volumes in the range of 0.003–0.007 $m^3$. $R^2$ values were 0.92, 0.84, and 0.77, while RMSE values were 0.00015 $m^3$, 0.00012 $m^3$, and 0.00008 $m^3$, respectively. The developed methods can be used to estimate the yield of cabbages under actual field and real-time harvesting conditions.

## 10. Intelligent crop production decision support with data fusion

Integration of data from various sources provides farmers with useful information and recommendations for managing their crops. Data fusion techniques can be used to combine data from different sources such as weather forecasts, satellite imagery, soil sensors, and crop growth models to create a holistic view of the crop's growing conditions. An intelligent system can use this information to provide farmers with real-time recommendations on things like irrigation, fertilization, pest control, and harvest timing. By using data fusion, the system can improve the accuracy of its predictions and recommendations, resulting in better crop yields and reduced costs for farmers. The data can also be analyzed by agronomists and researchers to improve crop management practices and develop new crop varieties. Toward this end, a system is being developed to fuse data from multiple sensors and multiple sites for two crops, namely pineapple in the open field and leafy vegetables planted indoors. It is hoped that the developed system can help farmers make more informed decisions, improve crop yields, and increase their profits while helping to conserve resources such as water and fertilizer.

## 11. Real-time sensor monitoring for greenhouse disaster resistance

Greenhouse agriculture mitigates external risks and ensures stable high-quality crop production. However, weather disasters frequently damage greenhouses, leading to significant economic losses. According to the Korea Meteorological Administration's Climate Change Assessment Report, the rate of change in climate factors (temperature) in Korea is twice the global average, expanding weather disasters such as drought, floods, and heat waves in the agricultural sector, and affecting crop quality, food safety, and productivity. The escalating impact of global warming and abnormal climatic events intensifies these damages. As the area of facility cultivation increases and cultivation techniques improve, ensuring the safety of facility structures becomes even more crucial. To minimize such losses, this study focuses on evaluating greenhouse safety and planning disaster-resistant structures. The research adopts lightweight and high-performance Inertial Measurement Unit (IMU) sensors to collect real-time data on rotation, displacement, and acceleration during diverse weather conditions. Automated data acquisition through these sensors is vital due to the challenges of manual monitoring. Because we have limitations of continuous data acquisition through personnel in the field. Utilizing open-source libraries, the study analyzes the dynamic data in three-dimensional space, enabling comprehensive insights into greenhouse behavior. Preliminary investigations assess the practicality and sensitivity of sensors under actual outdoor conditions. Moreover, the research compares greenhouse meteorological data with external weather station records, with a particular focus on wind speed and direction, significant factors in greenhouse damage. Understanding their impact enhances the resilience of greenhouse foundations. In conclusion, this study aims to fortify greenhouse agriculture against weather disasters through real-time sensor monitoring, contributing to long-term food security and agricultural competitiveness.

The most important sensor in this study is the nine-axis motion tracking sensor. In other words, it is an MPU6050 sensor called an inertial measuring device. The sensor incorporates a three-axis accelerometer, a three-axis gyroscope, and a three-axis geomagnetic system. The three-axis accelerometer represents acceleration, speed, and position data in a three-dimensional coordinate system. The three-axis gyroscope measures the angular velocity calculated by the Euler angle. It also supports an $I^2C$ serial interface called an inter-integrated circuit. It is very small in size, so it can make a miniaturized system. In addition, a microelectromechanical system technology called MEMS is applied. There is a 1024-byte FIFO buffer, which enables low power consumption. It also has a digital motion processor engine called DMP and a built-in thermometer capable of digital output. The microcontroller board, the core of the embedded system, used the WeMos D1 R1 board. This is an Arduino Uno-compatible board with a built-in Wi-Fi module called esp8266. It is a very small board with a length of 68 mm, a width of 53 mm, and a weight of 25 g. A Wi-Fi weather station was installed to collect weather data near the greenhouse. It collects information on indoor and outdoor temperatures and humidity, air pressure, UV levels, rainfall, wind speed, and wind direction. It sends data over Wi-Fi to store the data in a web browser.

In the case of gyroscopes, the MPU6050 has a drift phenomenon in which data are pushed over time within a low-frequency area. In the case of accelerometers, there is also a problem that a large number of noises are mixed in the high-frequency region by reacting sensitively to external vibration or environment. In order to increase the accuracy of low-cost sensors, a process of filtering data is essential. There are a wide variety of algorithms for the calibration of data. In this study, algorithms related to Complementary Filter and Kalman Filter were used. In addition, using Quaternion, an extended type of vector of complex numbers, the initial data of the IMU sensor was processed to obtain stable results. The DMP function of MPU6050 supplied by a company called Invensense was used. The value calculated by Quaternion in the FIFO buffer in the sensor is converted back to the Euler value to obtain Pitch, Roll, and Yaw values. As mentioned earlier, errors often occur between the low-frequency band data of the gyroscope and the high-frequency band data of the accelerometer. Therefore, the complementary filter extracts and corrects the posture angle using gyroscope data in the high-pass filter and acceleration data in the low-pass filter. In addition, the Kalman filter can correct data more accurately than the complementary filter, but the delay is large due to the complicated operation process, and it is too heavy to use in the MCU provided by Arduino. As a result, the data accuracy was improved by using the complementary filter and the MPU6050's DMP function, which has a simpler calculation process than the Kalman filter and allows smooth operation on a low-performance board (Arduino). Data were transmitted over Wi-Fi using a WeMos D1 R1 board with a built-in module called ESP8266. The transmitted data was stored in a web browser called Thingspeak and visualized as a graph. A weather station was installed directly to compare weather forecast data with weather data around the greenhouse. Like the embedded system board, a weather station capable of wireless Wi-Fi connection was used for data transmission. Graphs and data of weather data were stored in detail in units of days, months, and weeks.

Experiments were conducted mainly on wind speeds related to typhoons and strong winds, which have the greatest influence on weather phenomena that damage agricultural facilities. In this study, an experiment was conducted indoors at a wind speed lower than the standard of strong winds to determine the sensitivity and accuracy of the sensor. As the wind speed increases, the frequency and amplitude tend to increase almost proportionally, and in particular, there is a clear difference in the x-axis and z-axis. There was an error in which the amplitude instantly splashed, which was estimated to have been caused by a delay between data recognition or accumulation of data values as time accumulated in the process of processing data with a sensor. In the future, these sensors will be attached to the foundation of greenhouse structures to quickly detect the actual movement of the structure and use them to derive correlations by comparing and analyzing them with external weather conditions.

## 12. Boom height and pressure effects on spray droplet distribution

Excessive use of pesticides increases crop production costs and negatively impacts the environment. The efficiency of spray uniformity needs to be enhanced through the use of precision variable rate technology. The quality and effectiveness of the sprayer are greatly influenced by the technical performance of the nozzle. The objective of the research was to analyze the effects of operating pressure and height of the sprayer on the dispersal and uniformity of the spray droplets. To conduct the experiment, a test setup was constructed, consisting of four nozzles (NN D-35) and a single-cylinder motor with a four-stroke capacity of 0.72 kW. The applied speed was 2 km/h on the sprayer machine, while the test was conducted on the field platform. Experiments were conducted both in lab and field conditions with flat fan spraying nozzles and using water as the test liquid. The tested liquid outflow pressure ranged from 280 to 520 kPa. Depending on the spraying target surface, 35, 45, and 55 cm of working spray boom height were adjusted. The nozzle spacing was 30 cm, and the spray angle of the nozzles was 110°. In both laboratory and field conditions, the sprayer nozzle widths with boom heights of 35, 45, and 55 cm resulted in overlapping percentages of 22.38%, 23.43%, 24.76%, and 24.11%, 26.52%, and 29.59%, respectively. Under laboratory conditions, when the speed was set to 2 km/h, the average droplet density levels were 155.38, 159.20, and 171.91 spots/cm$^2$ at boom heights of 35, 45, and 55 cm, leading to spray coverage levels of 23.21%, 26.38%, and 28.35%, respectively. Similarly, in the field condition, at the same speed and boom heights, the average droplet densities were 138.62, 151.22, and 165.31 spots/cm$^2$, resulting in spray coverage levels of 23.88%, 25.61%, and 27.92%, respectively. Notably, the 55 cm boom demonstrated less than 30% overlap and offered optimal average droplet density and spray coverage per unit area in both scenarios.

## 13. Kinematic simulation of automatic pepper transplanter's seedling picker

The selection of an appropriate picking mechanism is crucial for the successful transplanting of seedlings. A clamp-type picking mechanism offers fast and efficient operations, reducing both the time and labor required for transplanting. The aim of this study was to conduct a kinematic simulation of a clamp-type picking mechanism for determining the optimal component dimensions. The clamp pin and bar picking device consisted of a manipulator with five grippers, a picking stand, gears, and a clamp bar. Among them, the picking stand, and manipulator with grippers dimensions are variables while the dimensions of gears and clamp are constant. A virtual simulation model with 10 trials of length combinations for the picking stand and manipulator with grippers was used to determine the kinematic characteristics such as position, velocity, and acceleration of the picking device variables to meet the trajectory requirements within the picking workspace and the azimuth angle of the seedling tray to prevent the seedling from breaking or falling. The length combinations considered range from 250 to 450 mm for the picking stand and 300 to 480 mm for the manipulator with grippers. Based on the kinematic simulation results, a 350-mm picking stand and a 380-mm manipulator were selected from the considered combinations. Under operating conditions of 30 rpm, the grippers exhibited maximum velocities of 0.63 m/s and 0.14 m/s and accelerations of 0.40 m/s$^2$ and 0.65 m/s$^2$ along the x-and y-axes, respectively at 900. The power required to complete the picking mechanism cycle was calculated to be 37.86 watts at a picking rate of 75 seedlings per minute. By assessing the suitability of the picking device's working trajectory, velocity, and acceleration of the grippers, a dimension suitable for the picking device was identified and will be validated with a field test. This study shows that by minimizing operating time, gripper velocity, acceleration, and damage to mechanical components, the picking system under development would improve pepper seedling picking accuracy and motion safety.

## 14. Proximal body temperature sensing in pigs using thermal imaging and deep learning

Pig body temperature is a crucial indicator of physiological and health status. Changes in pig body temperature can occur due to disease, different physiological stages, or external environmental conditions. Measuring pig body temperature allows comprehensive analysis of health conditions and physiological changes in growing conditions. Contact methods possess certain drawbacks for temperature detection, such as the difficulty of obtaining accurate and consistent readings, induce stress in pigs, and risk of contamination. Infrared thermography (IRT) is a non-contact measurement technology that can effectively measure pig body temperature without causing stress to the pig. A deep learning-based algorithm for the recognition and detection of pig body multi-parts temperatures using thermal images was proposed in this study. Data were collected in a lab-scale pig farm using a thermal camera with a camera angle of 45° from the top view. Videos were recorded from 4 pig sessions, and a total of 645 images were extracted with an interval of 25 frames from the video files. Then the data were split for training (580 images), validation (45 images), and test (20 images) datasets. A YOLOv5 model was trained to detect relevant pig body parts, followed by background segmentation to generate precise mean, median, and maximum temperatures for each detected body area. The results showed that the averaged precision and recall rate of the model for pig individual identification were 86.23% and 89.61%, respectively. The mean average precision (mAP) for body parts recognition was 91.32% when the threshold of the intersection over union (IOU) was set as 0.9. The study revealed that the lowest temperatures were observed on the pig nose, averaging 34.2°C, while higher temperatures were observed on the tail, ear, and head, with mean temperatures of 37.4°C, 36.8°C, and 36.9°C, respectively. The errors may be caused by surrounding temperature and humidity. Therefore, the proposed method showed optimal performance for pig body parts identification and temperature measurement in individual pigs provides a basis for automatic non-contact methods, and could be used for real-time temperature monitoring in the pig farm.

## 15. Automated greenhouse pesticide spraying with ultra-wideband technology

Greenhouse farming has gained considerable attention recently due to its role in ensuring food security and optimizing crop yields. However, the effective and correct application of pesticides in greenhouses remains challenging, including the potential risk to farmers from prolonged pesticide exposure [2]. This paper proposes a new method using Ultra-Wideband (UWB) [3] technology to develop an automated pesticide spraying vehicle for the greenhouse environment. The primary objective is to design and implement a UWB-based navigation and control system that enables the vehicle to move accurately throughout the greenhouse while also providing the capability to halt operations when encountering obstacles temporarily and automatically return to the refill station to resume interrupted spraying tasks. The pesticide spraying vehicle is entirely electric and powered by Po-Li batteries, ensuring continuous operation for a minimum of 2 hours to cover the entirety of the greenhouse for pesticide application. The wheels are equipped with high-traction rubber tracks suitable for the moist environment inside the greenhouse. Testing was conducted on a papaya field in Liu-Gui, Kaohsiung, Taiwan. The test results demonstrate the vehicle's precise adherence to the predefined path, successfully completing pesticide spraying across the entire 32 × 25 $m^2$ garden (**Figure 1**) [4].

The spray truck uses the UWB module for precise positioning in the greenhouse. Incorporate an additional IMU module to ensure accuracy when running straight and turning left and right in bends. All data are processed through a Jetson Nano computer. The map data, which stores the location information of the greenhouse needed for vehicle control, will be saved in a file and automatically loaded at system startup. The vehicle performs spraying according to the route from the starting position to the ending position. The vehicle uses a level sensor to measure the amount of liquid remaining in the tank. If water or pesticides run out during spraying, the car will save that position and automatically return to the refill position to add more liquid. After charging enough pesticide, the vehicle returns to the previous post and continues until the entire greenhouse area is sprayed. The test was conducted in the papaya garden in Liu-Gui Kaohsiung, Taiwan. As a result, the vehicle can complete the spraying stably and move accurately in the greenhouse area of 32 m × 25 m. Deviation from the center of the road is about ±20 cm but still within the allowable range. The time to complete spraying the entire area is 13 minutes. Tests for the vehicle's battery life are also tested separately. The car can operate continuously for 2 hours. The safety function that pauses when encountering an obstacle is tested and works correctly (**Figure 2**).

The research results demonstrate that developing an automated pesticide spraying vehicle for greenhouses using UWB technology offers a promising solution for precise and efficient pesticide spraying. Integrated UWB navigation, obstacle detection, and auto-return function ensure precise and uninterrupted spraying while reducing chemical use and minimizing farmer hazards. This advancement in greenhouse insecticide spraying technology has the potential to revolutionize agricultural practices, promote sustainable farming, and enhance crop protection strategies.

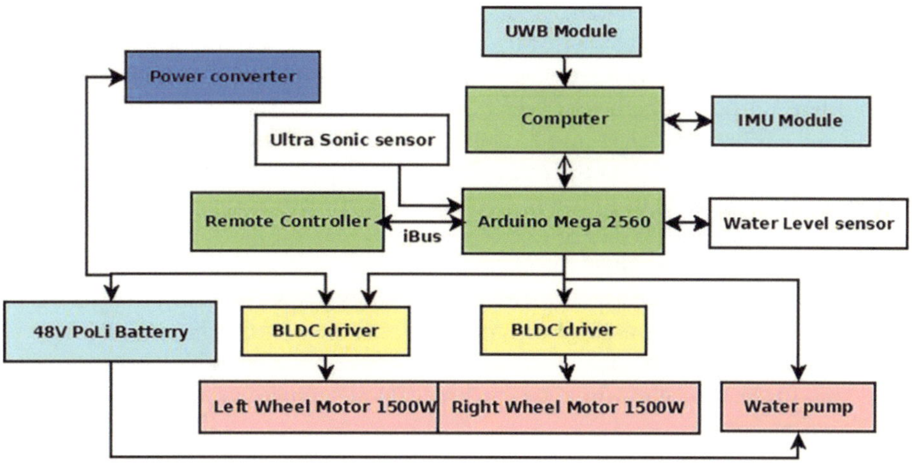


**Figure 1.**
*UWB-based car control diagram.*

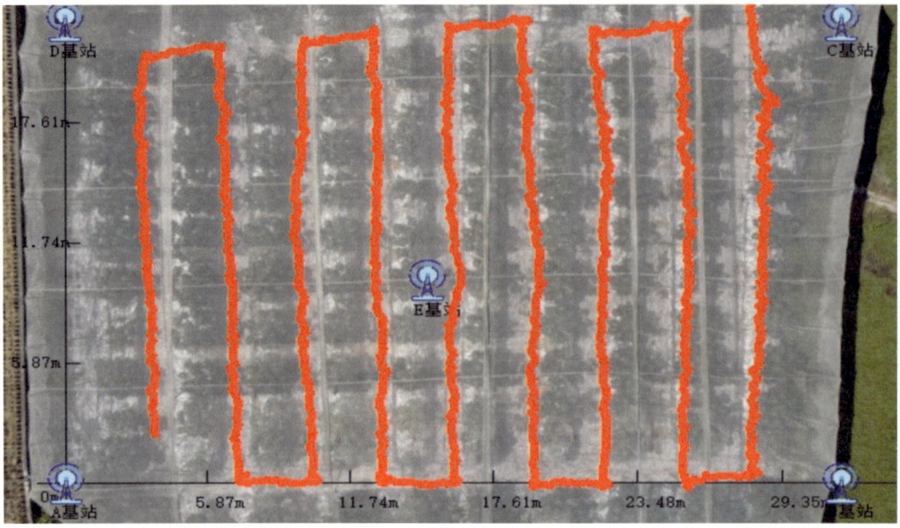


**Figure 2.**
*UWB-based car testing in papaya greenhouse.*

## 16. *Lonicera caerulea* fruits detection model

These days, digital twins, a digital model of an actual real-world physical asset, are considered for agriculture [5]. To achieve a digital twin for agriculture, there is a need to detect the spatial position of crops and their fruits. Drones are used for agriculture to sense crops [6] and to take images by drones [7, 8]. Deep learning is also used for the blueberry fruit detection algorithm [9], but often the position of the camera is close to the fruits, and the number of pixels of fruits was big. In this study, the aim is to evaluate the performance of a small (about 3–7 pixels to 3–10 pixels) *Lonicera caerulea* (the fruits similar to blueberry) detection model using deep learning. A drone camera (Mavic 3, DJI, China) was used to take *Lonicera caerulea* images. Images were taken in the *Lonicera caerulea* farm located in Chitose town in June 2023. The original image size was 5280 × 3956 pixels, but it was too big, so reduced the image size. Seventy-five images were annotated: the fruits of *Lonicera caerulea* were bordered as bounding boxes. Additionally, this image set was separated into 60 training and validation images and 15 test images. At the evaluation, precision, recall, and mean average precision of the test were calculated. The precision was 70.5%, and the recall was 71.5%. In addition, the mean average precision (mAP) was 65.5%. Regarding the images, the color of *Lonicera caerulea* was similar to shadows, and some predicted bounding boxes show the shade of *Lonicera caerulea* leaves. This study did image annotation, and the targets in images were only a few pixels to a few pixels. The result of detecting mini-size *Lonicera caerulea* fruits was 65.5% mean average precision. There is potential for tiny area object detection, even without no numerous image data. In the future, it would be used for digital twin research.

## 17. Automated spraying robot for cherry tomato greenhouse

The agricultural technology industry's demand for various materials and labor is extreme. Cherry tomatoes are grown in greenhouses using manual farming techniques, with labor representing the primary cost element and accounting for 41.6% of the overall production cost [10]. Taiwanese greenhouses employ manual spraying techniques like knapsack sprayers and automated irrigation systems. While automated spraying systems have significant installation and maintenance costs, using a knapsack sprayer harms users' health [11]. Consequently, introducing robots into this industry to automate specific farming tasks has been explored to optimize resource utilization and decrease the heavy dependence on human labor [12], ultimately leading to improved productivity and cost-effectiveness. During the spraying process, traditional manual spraying methods and even contemporary automated systems frequently ignore crucial elements like crop development phases and coverage density, leading to water waste and environmental contamination due to uncontrolled water application. By employing sensors [13] and camera [14] to identify the targeted irrigation regions, several attempts have been made to study how to maximize the irrigation water amount for crops. A line-based navigation system and an automated spraying system based on an estimate leaf area system (ELAS) have been combined to create an automated spraying robot (ASR). This research predominantly uses the ELAS technique to activate and deactivate relevant spray nozzles based on the determined leaf density zones. The utterly autonomous spraying vehicle may successfully use manual spraying techniques, lowering the danger to workers while providing a more consistent spray pattern and increasing the effectiveness of pesticide treatment in greenhouses for cherry tomatoes.

The control system comprises a microcontroller for the Vehicle Control System (VCS) and an embedded Spraying Control System (SCS) computer. The Arduino Mega 2560 controls the Autonomous Spraying Vehicle (ASR), receiving input from the remote and vision sensor, which is then transmitted to the vehicle's driver for precise navigation. To detect objects, a camera captures the scene, and the output is continuously processed by an embedded board (Jetson Nano) to identify the region of interest. The computer system utilizes ELAS results to optimize control for each nozzle valve. ELAS is developed to detect and measure leaf density in three regions of interest (ROIs). The image processing techniques use OpenCV for tree and leaf recognition through color-based segmentation. The workflow in **Figure 3c** involves transforming the image to the HSV color space, determining key parameters for object detection, and applying image noise filtering methods to enhance segmentation accuracy. The 50th percentile (Dv0.5) of droplet diameter was determined at various height positions used in field experiments (**Figure 4a**). There was no significant difference in droplet diameter distribution for the vertically sprayed ASR, with an average measured droplet diameter of 415.56 μm. According to the international standard S572.1 (2009), the spraying system of the autonomous vehicle produces a Medium to Very Coarse spray suitable for most agrochemical applications. The ASR demonstrated the ability to achieve more uniform and stable droplet distribution compared to traditional spraying methods using knapsack sprayers (**Figure 4b**). Results indicated no significant difference ($p > 0.05$) in Droplet diameter and droplet density between the two methods, suggesting that the ASR can achieve spraying efficiency comparable to manually operated sprayers. Due to the automation capability and lack of constraints faced by traditional methods, the ASR exhibited more than double the coverage compared to traditional spraying methods.

An ASR has been developed to distribute fertilizer and pesticides in greenhouses growing cherry tomatoes. A vertical spraying system with several nozzles and an ELAS based on machine vision may be used to distribute water droplets to specific leaf segments accurately and simultaneously. Due to accurate and reliable robotic

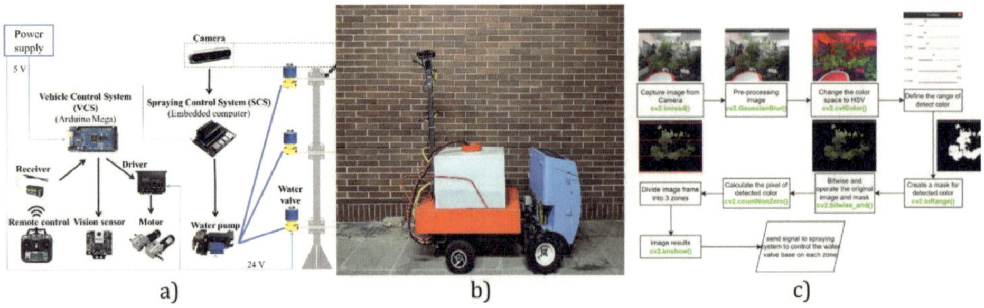


**Figure 3.**
*(a) System architecture, (b) prototype of ASR, and (c) block diagram of image processing.*

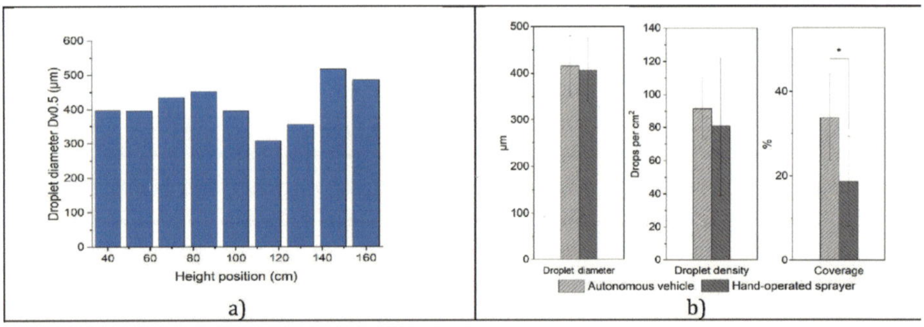


**Figure 4.**
*(a) Dv0.5 values (µm) per height position from 40 to 160 cm, and (b) comparing an ASR and a hand-operated sprayer. The symbol (*) indicates a statistical difference ($p \leq 0.05$).*

operations that minimize human error and encourage the best possible growth of cherry tomatoes, comparative studies reveal that the ASR surpasses manual spraying techniques. This technical development offers hope for successful and sustainable greenhouse farming. It will take advances in image processing methods to incorporate automated spraying systems with traditional sprayers.

## 18. Active back exoskeleton for rice farm lifting tasks

Lifting and carrying activities generally cause low back pain (LBP) and musculoskeletal disorders (MSDs) [15]. Farmers, including rice farmers, were eight times more prone to making significant modifications to their work activities because of LBP [16]. Despite squat-lifting being a safer lifting posture compared to stoop-lifting, more individuals tend to opt for stoop-lifting when lifting heavy objects, potentially due to its convenience and labor-saving benefits. Exoskeletons have recently emerged as a potential solution to alleviate these risks by reducing the physical strain on the human body during farming tasks [17]. A functional exoskeleton was effectively utilized, enabling the adjustment of assistive forces through actuators, regardless of the user's position. In contrast, passive exoskeletons lack the capability to adjust assistance or provide external power, relying solely on mechanical components such as springs and dampers. Due to the nonlinear nature of the dynamic system, which comprises both the user and the exoskeleton, a fuzzy controller proves to be appropriate for achieving torque response tailored to each back position [18]. This research aimed to design an active back exoskeleton to assist young rice farmers in their stand-lifting tasks. To regulate the exoskeleton's torque, a fuzzy controller was employed, utilizing feedback signals obtained from the Motion Processing Unit (MPU) sensor, precisely the angle estimated from the Accelerometer and Gyroscope data. By collecting and analyzing electromyography (EMG) data, we successfully assessed the benefits of using exoskeletons for lifting tasks among farmers.

The electromyography was utilized to monitor the muscle activities of the lower erector spinae (LES) and upper erector spinae (TES) during work-related movements as shown in **Figure 5**. The EMG signals from four muscles were captured using a portable EMG system (Nexus 10, Mind Media BV, Netherlands) at a sampling frequency of 2048 Hz. TES and LES EMG data were expressed as a percentage of a reference voluntary contraction (%RVC) to obtain a significant comparative measurement among participants. For this experiment, we employed two RMDx10 servo motors as actuators, each capable of delivering a maximum assist torque of up to 50 Nm. These motors can regulate the torque output by controlling the supplied amperage to the motors. In total, the weight of the exoskeleton amounts to 4.8 kg, which includes a 22.2 V 4.2 Ah lipo battery pack. The average operational duration on a single charge is approximately 2 hours. To control the motors, an Arduino Uno microcontroller is utilized, along with an MPU6050 sensor, to gather and estimate the back angle. Additionally, a CANbus module converts the SPI signal to Canbus, facilitating seamless communication between components. About the controller, we use the fuzzy controller to adapt the torque support correctly to the back angle. The controller consists of one input: the angle of the back; the output is the motor's torque. The input consists of seven trapezoidal membership functions. Similarly, the output also has seven trapezoidal membership functions.

In **Figure 6A**, the participants performed lifting motions without holding any weights in their hands. We observed that wearing the exoskeleton resulted in reduced

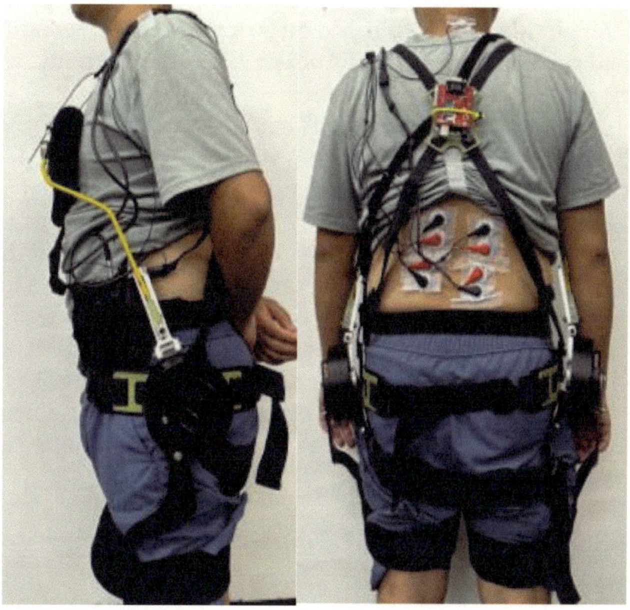

**Figure 5.**
*Illustration of the left-side and backside of the exoskeleton.*

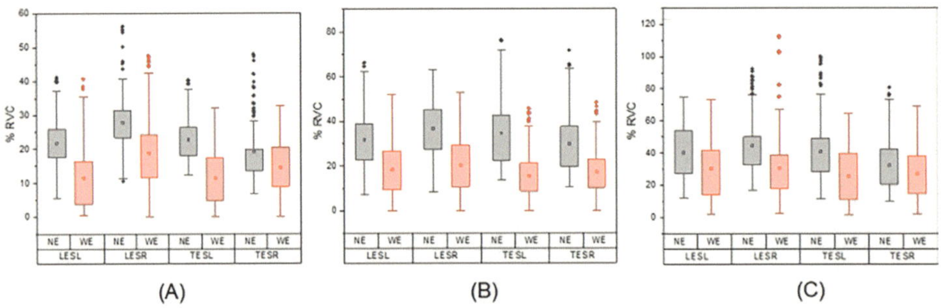


**Figure 6.**
*Box chart of participants' muscle activity, the lifting task without load (A), the muscle activity during lifting task with 5 kg load (B), the muscle activity during lifting task with 20 kg load (C).*

activation levels of the four muscles measured: the left-side TES muscle (TESL) decreased by 50.3%, the left-side LES muscle (LESL) decreased by 47%, the right-side TES muscle (TESR) decreased by 24.4%, and the right-side LES muscle (LESR) decreased by 32.6%. In **Figure 6B**, participants performed lifting tasks with a 5 kg load. We observed that wearing the exoskeleton resulted in reduced activation levels of the four muscles measured: TESL decreased by 48.7%, LESL decreased by 32.4%, TESR decreased by 30.8%, and LESR decreased by 32.6%. In **Figure 6C**, participants carried a 20 kg load while performing lifting tasks. We observed that wearing the exoskeleton led to decreased activation levels of the four muscles measured: TESL decreased by 36.9%, LESL decreased by 25%, TESR decreased by 16.6%, and LESR decreased by 31.2%. Providing support to users during lifting tasks is essential to minimize workplace-related risks. The use of exoskeletons leads to a reduction in muscle activation compared to lifting without them. This suggests that employing exoskeletons can help lower the risk of lower back pain (LBP), as assessed by the NIOSH lift equation. Moreover, the lifting equation is utilized to calculate the Recommended Weight Limit (RWL) based on task performance conditions, such as the horizontal and vertical position of the object relative to the body, vertical distance, frequency, and duration of the activity.

This research presents a novel active back exoskeleton that boasts a simple structure, lightweight design, and remarkable assistive capacity, all developed based on ergonomic design principles and biomechanical analysis methods that adapt the stood-lifting tasks. For control algorithms, we implemented a robust fuzzy adaptive control algorithm, enabling the actuators to provide torque in sync with the back angle, ensuring dynamic matching of assistive torque with changes in lifting load and the wearer's upper body's center of gravity position. Rigorous tests were conducted to evaluate the exoskeleton's performance, measuring reductions in EMG signals for both sides of LES and TES while lifting various loads ranging from 0 to 25 kg. As a result, this active back exoskeleton proves to be an effective solution for reducing the risk of wrist injuries and lower back pain during heavy object lifting tasks in rice farms.

## 19. Actuator failure detection for ICT-based orchard irrigation systems

Actuator in automated irrigation systems ensure irrigation scheduling and maintain optimum water content levels for better growth and yield of orchards. However, the event of actuator failures in automatic systems can lead to suboptimal irrigation practices and potential crop damage due to inadequate water distribution. The inherent complexity of actuator control in orchard irrigation systems necessitates effective troubleshooting to ensure optimal irrigation management. Appropriate and accurate detection of actuator failure is crucial for proper irrigation and water distribution. To address this issue, this study aimed to develop a machine learning algorithm (KNN) based methodology for detecting actuator failures for ICT-based orchard irrigation systems. A demonstration orchard consisting of four apple trees was established inside a greenhouse, with a soil test bench sized 3 m by 3 m. To enable time-based irrigation, the soil bin in the greenhouse was divided into two channels, where each channel accommodated two apple plants, allowing for separate irrigation schedules and management. The irrigation system comprised a single pump and two solenoid valves. The pump operated on a cycle of 60 seconds, alternating between ON and OFF states. The solenoid valves were operated on a cycle of 15 seconds, switching between ON and OFF states. These cycles are synchronized in different time phases over a period of 3 hours. Using a microcontroller, a Python program was coded to manage the irrigation pump and solenoid valves. This program facilitated the control of the irrigation components and storing and monitoring of all sensor data related to the irrigation system. Current sensors were employed to monitor the power consumption levels of the actuators, allowing for the assessment of their operational status. The current sensor signals were processed using Fourier analysis. By integrating data from control commands, current sensors, and water flow sensors, a KNN model was trained for the determination of the actuator status. The experimental results demonstrated that the proposed approach successfully detected the failure of solenoid valves, and the average power consumption was found to be 2.1 ± 0.02 W in the normal "ON" state, 0.024 ± 0.003 W in the normal "OFF" state, and 0.023 ± 0.003 W in the abnormal state, respectively. The proposed model achieved the highest accuracy of 93.6% for the value of k = 4 having the lowest mean error of 5% in detecting actuator failures. Our proposed model exhibited similar performance to comparable KNN architectures while utilizing fewer trainable parameters. A real-time fault detection and alerting feature will be added to KNN-based models in the future through internet-connected smartphone apps.

## 20. Water pressure influence on spatial and temporal soil water distribution

Effective water management is crucial for optimal crop irrigation in sandy soils where water pressure significantly influences water flow and distribution. Therefore, this study aims to investigate the effects of water pressure on spatial and temporal water distribution in sandy soils. The experiment was carried out in a plastic greenhouse facility at Chungnam National University, Republic of Korea. A subsurface drip irrigation system, with three rows (60 cm apart) and a 30 cm dripper interval in the pipeline, was designed and installed in a sandy soil testing bin (3 m × 3 m) inside the greenhouse for monitoring soil water movement and ambient conditions using soil water content, temperature, and humidity sensors. A total of 90 soil water content sensors were placed at different depths, ranging from 0 to 60 cm, to measure the flow of water in the sandy soil profile. Microcontrollers programmed with Python were utilized to communicate with the sensors, and data were collected for 24 hours per experiment and automatically saved as CSV files for further analysis. Different water pressures of 25, 50, and 75 kPa were applied at an emitter discharge rate of 3 L/h.

The average soil water contents at depths of 10, 20, 30, 40, and 50 cm varied from 49.428 ± 8.62% to 27.60 ± 5.23% during the experiment, with the ambient temperature and humidity ranging from 45 ± 3°C to 24 ± 3°C and 58 ± 2% to 50 ± 4%, respectively. The results showed that increased water pressure had important effects on the spatial distribution of water inside the soil bin. The increased pressure facilitated better infiltration and distribution of water, allowing it to reach deeper depths. However, at a 10 cm depth, the soil water content percentage was relatively higher compared to the content observed at a 50 cm depth, where it gradually decreased. The variations in water content percentages at different depths of sandy soil can be related to changes in ambient temperature and humidity and the limited soil water retention capacity of sandy soil. These findings would contribute to sustainable agricultural practices and enhance the understanding of water movement in sandy soil, thus promoting efficient water use and crop growth.

## 21. Theoretical analysis of automatic pepper transplanter's mechanism

Mechanized transplanting of pepper seedlings is an efficient and labor-saving method. The type of planting mechanism used in the transplanter has a significant impact on overall efficiency. Theoretical analysis allows for the determination of the kinematic parameters of transplanting components, such as appropriate dimensions, operating speed, power requirement, plant spacing, and device strength, for effective transplantation. The goal of this study was to conduct a theoretical analysis of a four-bar link-type transplanting mechanism using 3D modeling in order to determine the optimal seedling planting speed and interval. The planting trajectory and depth were considered in order to maintain inter-row distance and planting interval, which were influenced by both crank motion and transplanter forward speed. The link-type transplanting mechanism consisted of three major components: a crank, a coupler, and a transplanting hopper. Five kinematic simulation trials of crank length, trajectory height, and width were performed in static conditions to determine the appropriate crank and planting trajectory dimensions. To select the optimal operating speed and planting spacing, five kinematic simulation trials of forward speed, velocity, and acceleration were performed in dynamic conditions. To maintain a planting depth of 120 mm, the appropriate crank length, trajectory height, and width were determined to be 60 mm, 254.57 mm, and 106.05 mm, respectively. The appropriate transplanter forward speed was determined to be 0.15 m/s for a rate of 60 seedlings per minute. With a planting spacing of 150 mm, the velocity and acceleration along the x- and y-axes were 0.393 m/s, 0.950 m/s, 4.026 $m/s^2$, and 7.048 $m/s^2$, respectively. The power required to plant one seedling for the link-driven hopper-type dibbling mechanism made of steel alloy 1020 was calculated to be 11.65 W. The findings of this study would be useful for improving the link-type dibbling mechanism design, and the validation would require field tests.

## 22. Stress classification in controlled pepper seedling growth using image analysis and machine learning

Environmental stress impacts plant growth adversely under controlled environments, where crucial factors such as temperature, humidity, light intensity, and $CO_2$ levels can profoundly affect plant development. These stresses exert direct effects on seedlings, leading to changes in leaf size, leaf count, overall plant size, and canopy area. Manual inspections are time-consuming, promoting the analysis of morphological features such as the shape, color, and texture of pepper seedlings for stress quantification. Thus, this study focuses on extracting features from images to classify pepper seedlings as stressed or non-stressed across four distinct environmental conditions. In a controlled plant factory, 2-week-old pepper seedlings were subjected to different temperatures (20, 25, and 30°C), light intensity levels (50, 250, and 450 PPFD), water supply levels (1, 2, and 3 L/day), and day and night hours (8/16, 10/14, and 16/8). Fifteen images of the seedling canopy were captured daily from each stress using a low-cost RGB camera over a 2-week period, resulting in a dataset of 2520 images from the five chambers of the plant factory. Color and texture features were extracted from each image using the gray-level co-occurrence matrix (GLCM). Statistical analyses, including the two-way analysis of variance, were performed to assess the significance of the treatment effects. Feature selection algorithms, including Duncan's multiple range test, dominant color descriptor (DCD), color layout descriptor (CLD), and sequential forward floating selection (SFFS), were employed to reduce feature overlaps. Four classification models, such as support vector machine (SVM), naive Bayes classifier (NBC), K-nearest neighbor (KNN), and random forest classifier (RFC), were utilized to accurately classify the stress and non-stress states of pepper plants. The impact of stress on seedlings in the plant factory was observed after approximately 3 days of their placement. Morphological analysis revealed significant differences in stress conditions caused by temperature variations and water content levels. However, light intensity and night hours did not show significant differences at a 5% significance level and Duncan's multiple range test analysis. The SVM, KNN, NBC, and RFC classifiers achieved average stress classification accuracies of 90.03%, 84.44%, 77.34%, and 61.5%, respectively. These findings provide valuable insights for real-time stress monitoring and optimizing environmental conditions for enhancing growth and effective monitoring in controlled cultivation.

## 23. Multispectral image enhancement techniques for vegetation index calculation

Multispectral imagery plays a crucial role in vegetation analysis, particularly in calculating the normalized difference vegetation index (NDVI). The field of image data enhancement continues to hold significant interest due to challenges such as atmospheric interference, sensor noise, and uneven illumination, all of which can adversely affect the quality of raw data. However, the effectiveness of image data enhancement techniques in improving NDVI accuracy remains an area of interest. This study aimed to evaluate the performance of field of view (FOV) alignment and histogram equalization methods as an enhancement technique for multispectral imagery in vegetation indices (VI) calculation. A multispectral sensor (MicaSense Rededge MX) was employed to capture raw imagery, while an active canopy sensor (Crop-circle, ACS-435) collected the reference data. Normalized differential vegetation index (NDVI), the most extensively practiced VI, was chosen for the analysis. Data were collected from a wheat field at four different growth stages (GS1: 10, GS2: 34, GS3: 70, and GS4: 84 days after sowing). The MicaSense, Crop-circle, and a GPS receiver were mounted on a handheld structure, and data were taken from a constant height of 90 cm from the canopy. GPS locations and pixel-based image segmentation, resembling the Crop-circle FOV was processed using a Python language-based program. NDVI was calculated using Red and NIR reflectance data extracted from the cropped portions of the images. Histogram equalization ensured balanced pixel intensity distribution. NDVI was calculated from the original images without FOV segmentation and histogram equalization, and data enhancement techniques were evaluated using the regression analysis for the assessment. Results indicated that the coefficient of determination ($R^2$) and root mean square error (RMSE) between the reference and raw image value were (0.18, 0.1163), (0.17, 0.4017), (0.41, 0.3289), and (0.82, 0.2104) on GS1, GS2, GS3, and GS4, respectively. For the enhanced image dataset, the $R^2$ and RMSE values exhibited GS1 (0.61, 0.0768), GS2 (0.69, 0.3121), GS3 (0.56, 0.2902), and GS4 (0.85, 0.1619), respectively. GS1 (43%) and GS2 (42%) provided significant accuracy improvements which showed the effectiveness of FOV alignment and histogram equalization techniques. In the early growth stages, plants experienced a sparse canopy with less distinct spectral signals affected by noise and atmospheric interference, resulting in lower $R^2$ values and higher RMSE. However, enhancement techniques like FOV alignment and histogram equalization effectively mitigate errors by refining spatial alignment, noise reduction, and image contrast enhancement. Such enhancements empower researchers and practitioners to make reliable decisions, particularly in crop monitoring and management during initial development stages.

## 24. Biodegradable potted seedling transplanter

Plastic seedling pots have been widely used due to their lightweight and durable nature, but they hinder root establishment efficiency. However, recent studies have demonstrated that biodegradable potted seedlings could improve seedling resilience, while also being eco-friendly through natural decomposition. In this research, a transplanting mechanism for biodegradable potted vegetable seedlings was designed using commercial software, incorporating kinematic analysis to enhance the efficiency and performance of leafy vegetable transplantation. Theoretical analysis of the vegetable transplanting mechanism for biodegradable seedling pots was conducted, including calculations of position, velocity, acceleration, and input driving torque. Additionally, the selection of appropriate link combinations within the mechanism was explored to ensure the smooth transplantation of potted seedlings in optimum depths and spacings. The kinematic model of the transplanting mechanism was simulated using commercial mechanical design and simulation software. The transplanter was comprised of a 4-bar mechanism: a driving link, a driven link, a connecting link, and a supporting bar. To enable better hopper motion, a spring was affixed between the driven link and the ground. The movement of the mechanism was primarily controlled through a crank-rocker mechanism, where the arm lengths play a crucial role in determining the planting trajectory. Simulation trials were conducted by varying the main arm link length while maintaining fixed forward and rotating speeds. The following parameters were investigated: driving link (45–65 mm), connecting arm (130–170 mm), guide bar (115–135 mm), and end effector link (210–250 mm). Additionally, a dibbling hopper length of 153 mm was identified as ideal. The simulated velocities and accelerations of the end hopper in "X" and "Y" directions for suitable link combination were found to be 430 mm/s, 530 mm/s, and 975 mm/s$^2$, 2091 mm/s$^2$, respectively. The required driving torque was observed to be 603 N mm, and the vertical linear displacement of the hopper was 281 mm.

## 25. Distributed sensing for agricultural mobile robot navigation

Autonomous navigation of agricultural mobile robots through uneven terrains, avoiding obstacles, and optimizing routes are intricate challenges that demand innovative solutions. Conventional navigation methods, often reliant on GPS alone, face limitations in accuracy and reliability, especially in densely vegetated areas with poor satellite reception and disturbances of the outdoor environment. Data fusion and multiple perception solutions are usually employed to assist the existing RTK GPS-based navigations and to improve the reliability of the operation in unstructured agricultural fields. In this context, a distributed sensing system (DSS) that leverages multiple data sources can mitigate these limitations and maintain reliable navigation even in challenging conditions. This study reports on the design, development, and evaluation of a DSS with CANBUS communication to assist the autonomous navigation of a four-wheel steering agricultural mobile robot inside berry orchards. The proposed DSS was responsible for detecting random obstacles and providing the navigation controller with real-time feedback for effective collision avoidance. For the proof of concept, the assisted navigation was expected to maintain the robot between the plants' rows with an accuracy of 10–20 cm from the side with an ideal forward speed of 5–8 km/h. In the hardware layer, the DSS benefits from a Jetson Nano onboard computer along with a set of ROS-based multi-channel infrared and laser sensors for perception. In the software layer, a fuzzy knowledge-based algorithm was designed and simulated. Results demonstrated significant improvements in navigation accuracy through the fusion of GPS, laser, and infrared sensor data which was essential to avoid crop damage and ensure optimal resource utilization. Results from simulation and field experiments also suggested that the exponential filter was necessary to be implemented on each sensor for removing noise and outliers. It was concluded that the development of such systems requires extensive validation tests in different orchards besides a more accurate dynamic model of the robot platform. This can be accelerated using digital models inside virtual replicas of the environments.

The base vehicle was a combustion engine four-wheel drive (4WD) and four-wheel steering (4WS) Quatrak manufactured by Irus (IRUS Motorgeräte GmbH, Burladingen, Germany) with a track width of 1.2 m, maximum forward and reverse speed of 10 km/h, and approximate weight of 475 kg that utilizes a 2-cylinder petrol engine with 20.1 kW (27 HP) output power. The 4WS mechanism of the vehicle allowed a minimum turning radius on the row-ends with increased precision and control over the robot's movement, which is of great importance for making sharp turns and navigating in tight spaces inside orchards. The vehicle benefits from the HYDAC TTC 580 (HYDAC, Sulzbach, Germany) controller to translate and deliver the control steering signals to the actuators (including the electrohydraulic system) that are responsible for speed and turning angle. A GPS-based navigation toolbox was installed on the vehicle by Innok Robotics (Innok Robotics GmbH, Regenstauf, Germany) that benefits from a custom-built graphical user interface for uploading waypoints and following trajectory. To validate the proposed assisted navigation system, field visits were first carried out to collect preliminary data using high-accuracy RTK GPS. These data were used to create a virtual orchard inside CoppeliaSim that was interfaced with ROS for testing different sensors, hardware in the loop, and control algorithms on a full-scale simulated robot and orchard model. The main elements of the simulation scenes in this project were (i) mesh files representing plants, the robot, and obstacles, (ii) API and codes that created interfaces between different software environments, and (iii) algorithms and dynamic models including minimum distance calculation, Ackerman steering, path following, and obstacle avoidance algorithms. The simulation approach converted native data streams from various sensor inputs into usable information within the command-and-control system. The sensors used in front of the robot were TFmini Plus single-point short-range LiDAR sensor manufactured by Benewake (Benewake, China) with a distance detection range between 0.02 and 12 m, resolution of 1 cm, sampling rate of 100 Hz, frame rate between 1 and 1000 Hz, accuracy of ±1%, and field of view of 3.6°. In the communication layer, CANBUS (ISO 11898-2) was used for exchanging data between different nodes due to its scalability and high reliability for automation even in harsh environments with high electromagnetic interference.

The simulation scene provided a safe, fast, and low-cost experiment platform for developing, testing, and validating of the sensing and control strategies with different algorithms. It also enabled human-aware navigation by finding the best positions for each sensor and provided a flexible solution for attaching other implements and determining the optimum row-end turning patterns in the presence of random obstacles. It also accelerated complicated analysis to understand the behavior of the robot on uneven terrains. Results of field experiments showed the reliable functionality of these sensor arrays under outdoor light conditions to detect the bushes and random obstacles within the pre-programmed range of 5–60 cm. In addition to the detection task, each sensor module also transmits a state signal equal to 0 or 1 on the CANBUS line in the absence or presence of an obstacle that is georeferenced with the robot position for monitoring purposes. The average distance between the robot and the left bushes, as measured by the three sets of IR sensors mounted on the left of the robot were 17.5, 18.5, and 17.1 cm. These values for the right bushes were found to be 13.5, 18.9, and 16.6 cm. It was observed that the front IR sensor provided the collision avoidance controller with accurate feedback to stop the robot nearly 25 cm distance from the bushes. These results show the redundancy in the distance measurements that the multi-channel IR sensing setup is providing to make sure the controller always receives feedback, and the robot continues its operation, even if one sensor module fails. Field experiments also confirmed that the analog output of these sensors with the exponential filter noise reduction allows for precise measurements, even in high-density bush conditions, and the multi-channel ensured that the final measurements were not easily affected by the disturbances of the environment such as occlusion by leaves and dust. From the perspective of exchanging sensor data with the controller, the CANBUS network was found to be robust and reliable with 100% of the data transmitted during the experiment without any interruption.

The system showcased consistent performance across different types of crops, soil conditions, and environmental factors. The hardware of the system was designed to be modular and could be replaced easily upon failure. This adaptability stems from the fusion of multi-modal sensor data, which allows the robot to gather comprehensive information and make informed decisions regardless of the surroundings. Comparative analysis with existing navigation methods highlighted the superiority of the DSS. The quantitative comparisons demonstrated the system's significant advantages in terms of accuracy, efficiency, and adaptability. The system's ability to generalize its navigation strategies to various contexts holds great promise for its widespread adoption in agriculture. The evaluation confirmed that the proposed solution is capable of adjusting the robot's steering and speed steering the robot in the presence of different obstacles and different scenarios. This was achieved by a control strategy that reduced the error distance between the sensor readings and a minimum allowed distance with left, right, and front obstacles.

## 26. Key technologies for intelligent kiwifruit production

Kiwifruit is popular all over the world because of its unique taste and high nutritional value. Traditional kiwifruit production approach almost relies on manual operations, which is not only labor-intensive but also easily affected by human factors, resulting in low production efficiency and unstable output. As the demand for kiwifruit continues to rise, inefficient manual operation is highly desired to be replaced by intelligent production with advanced technologies. Precision operation robots supported by artificial intelligence and automation technology have been proposed and made contribute to improving agriculture production efficiency and reducing labor usage. Therefore, this research is focused on key technologies for fully intelligent production of kiwifruit, with a view to replacing manual operations, reducing labor dependence, and improving overall productivity in bud thing, flower pollination, yield estimation, and robotic picking. Buds of kiwifruit are usually grown in clusters, consisting of a main bud and multiple side buds that grow around it, as shown in **Figure 7**. As an important part of nutrient management during the flowering period, bud thing refers to the removal redundant side buds that are not suitable for fruiting, which directly affects final yield and quality. An automated robotic precision bud thing robot based on machine vision and laser was developed to automatic detect side buds and destroy them to prevent their growth.

The key to the proposed method is to accurately and quickly detect side buds that need to be thinned out in the canopy. For this regard, a multi-class buds labeling, and detection strategy was proposed according to its different growth conditions and spatial location, as shown in **Figure 8**. All collected kiwifruit bud image dataset was labeled with the mentioned strategy, and YOLOv5l was trained and employed as bud detection model. Apart from this, a fixed-focus laser was selected as bud thing actuator for robots to destroy side buds by emitting high energy laser pulses.

As a typical cross-pollinated plant, kiwifruit does not have the ability to achieve autonomous self-pollination. Due to the strong subjectivity of manual work, artificial assistance pollination approaches have problems such as uncontrollable pollination, and even missed pollination, which affects fruit quality and yield. Robotic kiwifruit pollination has been proposed for saving people from heavy labor and reducing pollen losses by imprecise pollination in orchards. One of the key technologies for precision pollination of kiwifruit is the reliable detection of flowers suitable for pollination. As the flowering period of kiwifruit is not synchronized, not all flowers are blooming and able to receive pollen at the same time [19]. Therefore, a multi-class flower detection method (shown in **Figure 9**) based on YOLOv5l was employed for detecting and determining which flowers

**Figure 7.**
*Example of bud clusters with a main bud and surrounding side buds.*

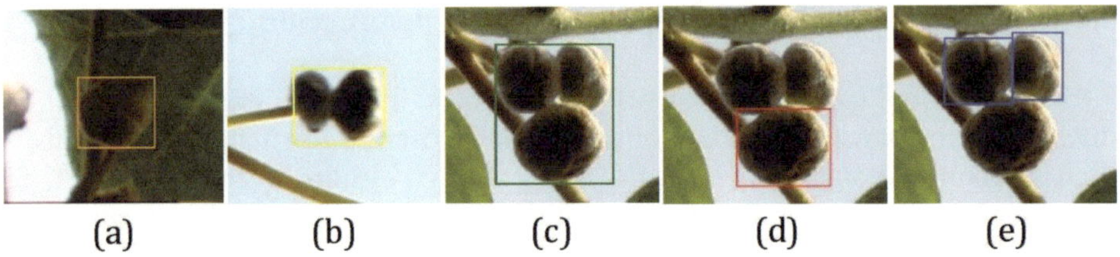


**Figure 8.**
*Labeling examples of multi-class kiwifruit buds with different colors. (a) Single bud; (b) double buds, (c) bud cluster, (d) main bud, (e) side buds.*

**Figure 9.**
*Labeling examples of multi-class kiwifruit flowers with different colors. (a) Bud; (b) early open; (c) half-open; (d) fresh pistil; (e) early ocher pistil; (f) ocher pistil; (g) petal fall; (h) occluded pistil of flower; (i) bright pollen; (j) dark pollen.*

of the canopy were in the best pollination periods. After that, a further selection strategy based on Euclidean distance matching method was applied to obtain its distribution in the canopy for suitable flowers selection, which combined the agronomic characteristics of kiwifruit growth for optimal nutrients partition with quality and yield assurance [20].

Pollination approach is another essential consideration for kiwifruit pollination robots to achieve precise operations. An air assisted liquid pollination approach was designed and built to collaborate with a robotic arm for kiwifruit flowers targeting pollination, which also could achieve quantified pollination by controlling the spraying time of prepared pollen liquid [21]. Comprehensive consideration of flower size and effective pollination area, a pre-experiment was carried out for pollination parameters optimization and selection, which determined parameters of the distance between nozzle and the selected flower was 25 cm, air pressure was 56 kPa, and liquid flow was 45 mL/min, respectively. Fruit yield estimation before the harvest is a crucial step to predict the required resources for workers such as packing and storage houses for the harvest and distribution resources for the marketing. For requirements of rapid yield estimation in small and medium-sized kiwifruit orchard, single shot multibox detector (SSD) with two lightweight backbones MobileNetV2 and InceptionV3 were employed to develop an Android APP for field kiwifruit detection [22]. However, difficulty in qualifying the planting area from smartphone acquired images and video frames makes yield estimates less credible. Therefore, a multi-target tracking and counting algorithm based on the trained YOLOv5 kiwifruit detection model and the ByteTrack tracking framework is proposed, which achieves accurate kiwifruit counting by restricting the single-row region based on the column positions fed back from the detection model, as shown in **Figure 10**.

Kiwifruits are commercially grown on sturdy support structures such as T-bars and pergolas. Automatic detection of kiwifruit in the orchard is challenging because illumination varies through the day and night and because of color similarity between kiwifruit and the complex background of leaves, branches, and stems [23]. Also, kiwifruits grow in clusters, which may result in having occluded and touching fruits [24]. Intelligent sensing and nondestructive picking of fruit are the two main key technologies for robotic harvesting. One of the hardest challenges in orchard automation is the harvesting robot especially for soft and delicate fruits. However, all the kiwifruits have been labeled and detected as only one class in most research for fruit robotic picking, where fruits occluded by branches or wires have been detected as pickable targets. End-effectors or robots may be damaged by the branches or wires when they are forced to pick those fruits [25]. Therefore, kiwifruits are labeled, trained, and detected in multi-classes based on their occlusions to avoid detecting fruits occluded by branches or wires as pickable targets [24]. Fruits are classified into five classes according to robotic picking strategy and field occlusions, as shown in **Figure 11**. Well-known YOLOv4 was employed to do transfer learning for multi-classes kiwifruit detection.

Fruit nondestructive picking is the other key technology for developing picking robots. Firstly, based on the artificial way of kiwifruit picking and the biological characteristics of the kiwifruit stem, a fruit-picking method for the robot was proposed, which needed to separate the fruit from the stem and hold the fruit to prevent it from dropping. Then, the picking method was verified by a specially designed separation test of the fruit and its stem. After that, an end-effector was designed and manufactured based on

**Figure 10.**
*Kiwifruit yield estimation based on a multi-target tracking and counting algorithm.*

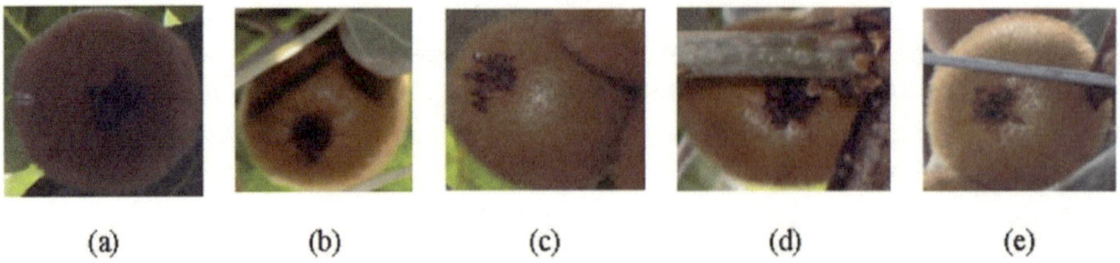


**Figure 11.**
*Different classes of kiwifruit images. (a) Fruit not occluded (NO); (b) fruit occluded by leaves (OL); (c) fruit occluded by other fruits (OF); (d) fruit occluded by branches (OB); (e) fruit occluded by wires (OW).*

the fruit-picking method, which approached a fruit from the bottom, enveloped and grabbed the fruit from two sides, and then rotated up to separate the fruit from the stem [26]. Furthermore, in order to investigate the drop distance of the kiwifruit end-effector to container without damage, a low-damage crash study on kiwifruit was conducted [27].

The above key technologies were integrated and verified in standardized commercial kiwifruit orchards. According to the agronomic analysis of different production requirements, a comprehensive operating robot with a mechanical arm as the main structure was determined and constructed. For this robot, it is only necessary to replace different end-effectors and corresponding control algorithms to achieve different tasks switch between bud thing, flower pollination, and robotic picking, as shown in **Figure 12**. Field experiments have shown that the average time for a constructed multifunctional robot to thin a bud, pollinate a flower, and pick a kiwifruit is 1.5 s, 1 s, and 2 s, respectively. For kiwifruit yield estimation, 90% of counting accuracy can be achieved in video frames with less jitter.

Precision and sustainable intelligent production of agriculture are desired to meet the needs of the growing world population. This research proposed key technologies for the fully intelligent production of kiwifruit, focusing on bud thing, flower pollination, yield estimation, and robotic picking. By analyzing the agronomic characteristics of

**Figure 12.**
*Constructed multifunctional precision operation robot. (a) Bud thing; (b) flower pollination; (c) robotic picking.*

different tasks, a general-purposed precision operation robot was designed and adopted for kiwifruit intelligent production with a mechanical arm structure, which could flexibly switch between different tasks of bud thing, flower pollination, and fruit-picking by replacing different end effector and control algorithm. In addition, accurate preharvest yield estimation provides data support for labor assessment. However, the promotion of intelligent agricultural technologies still faces several challenges, such as equipment maintenance costs and the acceptance of new technologies by farmers. Therefore, future research can continue to explore more economically practical intelligent equipment, strengthen the training and awareness of farmers on the new technology, and promote the application of intelligent agricultural technologies in a wider range.

## 27. UAV imaging for estimating pineapple crop growth

Hormonal induction is an activity performed on pineapple plants to control flowering time or breaking the vegetative phase into generative using ethylene [28]. Based on farmers' experience, ethylene is usually sprayed on pineapple plants that are 7–10 months old after planting. This practice is performed without considering the level of crop growth. Hormonal induction done on crops that are not mature enough may produce small fruit or may not produce fruit at all. On the other hand, a delay in the induction of hormones will produce over-size fruit. Fruit sizes that are too large have less export demand due to packaging and storage space limitations. Therefore, it is important to accurately estimate the hormone induction time-based on the crop growth stage so as to meet the local and foreign markets' demand. The crop growth stage is determined based on the crop characteristics that include the length of the longest leaf (the D-leaf), the plant height, and the number of fresh leaves that emerge from each plant [29, 30]. A study by Usman et al. [29] reported that plant weight is an indicator parameter in determining the crop growth stage. This is done by removing several plant samples and weighing them. The manual practice of measuring these parameters is, however, laborious, time-consuming, tedious, and limited to several numbers of samples for large plantation areas. Moreover, the method of weighing some plant samples also reduces the yield because the plants do not grow well if replanted. The direct contact between operators and crops during the growth parameter measurement may also expose the crop to disease transmission [31]. A nondestructive method that can be done remotely is therefore required to replace the manual method of crop growth estimation, especially for large plantation areas. For this purpose, a study has been conducted to investigate the potential of the unmanned aerial imaging system (UAIS) for estimating pineapple crop growth.

The UAIS comprised a drone, a multispectral camera, and a control module. The multispectral camera that was mounted on a DJI Phantom 4 drone consists of six sensor bands namely visible (RGB), Red Edge (RE), Near-infrared (NIR), Red (R), Green (G), and Blue (B) with the resolution of each sensor being 2 MP. The drone was flown at 40 m from the ground with autopilot mode to capture multispectral images of pineapple crops at all six bands. The precise position location up to the centimeter level for each image taken was also acquired using the built-in global positioning system (GPS). All the captured images were then stitched and processed using Pix4Dmapper software to form an ortho-mosaicked image. The ortho-mosaicked image was further analyzed and formulated using Quantum GIS (QGIS) software to form vegetation indexes that include normalized difference vegetation index (NDVI), normalized difference red edge index (NDRE), and ratio vegetation index (RVI). Measurements of two pineapple crop growth parameters such as plant height

| Crop growth parameter | Correlation coefficient (R) | | |
|---|---|---|---|
| | NDVI | NDRE | RVI |
| Plant height (cm) | 0.88 | 0.86 | 0.83 |
| D-leaf length (cm) | 0.90 | 0.87 | 0.84 |

**Table 1.**
*Correlation coefficient between the average crop growth parameters with the vegetation indexes.*

and D-leaf length were then carried out on eight plant samples from each of the 24 subplots. The average value of each parameter for eight samples from each subplot was calculated to obtain one representative growth parameter value for each subplot. The average values of the crop growth parameters obtained were then correlated with the average values of the vegetation indexes obtained from the multispectral images.

The results showed that among the three vegetation indices, NDVI attained the highest correlation with plant height and leaf length-D, with correlation coefficient (R) were 0.88 and 0.90, respectively. This is followed by NDRE with R values of 0.86 and 0.87 for plant height and D-leaf length. The lowest correlation is the RVI index with R values for plant height and D-leaf length of 0.83 and 0.84 respectively. These results demonstrated that multispectral images taken using UAIS can be used to estimate the growth rate of the pineapple crop. **Table 1** shows the correlation coefficient between the average crop growth parameters with the vegetation indexes.

The use of UAIS for estimating pineapple crop growth rate has been reported briefly in this paper. The good correlation between the crop growth parameters with the vegetation index formulated from the multispectral image taken using UAIS indicates that the UAIS can be potentially used to replace the manual practice of crop growth parameter measurement. The crop growth stage information can be uploaded into a farm management system based on the Internet-of-Things (IoT) which then can be used as a reference for farm managers in making agronomic decisions.

## 28. Conclusion

Emerging technologies in precision agriculture have demonstrated promising results in revolutionizing the industry. Through the integration of machine learning, robotics, IoT sensors, and drones, precision agriculture offers unique opportunities to optimize resource utilization, enhance crop yields, and mitigate environmental risks. The abstracts presented in this chapter showcased that the convergence of technology and agriculture not only increases operational efficiency and profitability but also promotes sustainability and resilience in the face of challenges such as climate change and food security. Advancements such as actuator failure detection for ICT-based orchard irrigation systems and the implementation of active back exoskeletons for rice farm lifting tasks highlight the industry's commitment to efficiency and sustainability. Additionally, research into water pressure influence on spatial and temporal soil water distribution and the theoretical analysis of automatic pepper transplanter mechanisms underscore the importance of optimizing resource utilization and minimizing environmental impact. By fundamentally reshaping traditional farming methods, these innovations pave the way for sustainable and efficient practices within evolving challenges and uncertainties, providing insights into the future of agriculture, where data-driven decision-making and smart technologies play a central role in addressing the complex challenges facing the global food system.

## Author details

Redmond R. Shamshiri[1*], Maryam Behjati[2], Siva K. Balasundram[3], Christopher Teh Boon Sung[3], Ibrahim A. Hameed[4], Ahmad Kamil Zolkafli[5], An Ho-Song[6], Arina Mohd Noh[5], Badril Hisham Abu Bakar[5], W.A. Balogun[7], Beom-Sun Kang[8], Cong-Chuan Pham[9], Dang Khanh Linh Le[9], Dong Hee Noh[10], Dongseok Kim[11], Eliezel Habineza[6], Farizal Kamaroddin[7], Gookhwan Kim[12], Heetae Kim[12], Hyunjung Hwang[11], Jaesung Park[11], Jisu Song[11], Joonjea Sung[13], Jusnaini Muslimin[14], Ka Young Lee[6], Kayoung Lee[6], Keong Do Lee[12], Keshinro Kazeem Kolawole[7], Kyeong Il Park[15], Longsheng Fu[16], Md Ashrafuzzaman Gulandaz[6], Md Asrakul Haque[17], Md Nasim Reza[17], Md Razob Ali[17], Md Rejaul Karim[17], Md Sazzadul Kabir[6], Md Shaha Nur Kabir[18], Minho Song[15], Mohamad Shukri Zainal Abidin[7], Mohammad Ali[17], Mohd Aufa Md Bookeri[5], Mohd Nadzim Nordin[5], Mohd Nadzri Md Reba[19], Mohd Nizam Zubir[5], Mohd Saiful Azimi Mahmud[7], Mohd Taufik Ahmad[5], Muhammad Hariz Musa[5], Muhammad Sharul Azwan Ramli[20], Musa Mohd Mokji[7], Naoto Yoshimoto[21], Nhu Tuong An Nguyen[9], Nur Khalidah Zakaria[22], Prince Kumar[23], P.K. Garg[23], Ramlan Ismail[5], Ren Kondo[21], Ryuta Kojo[21], Samsuzzaman[17], Seokcheol Yu[12], Seok-Ho Park[12], Shahriar Ahmed[17], Siti Noor Aliah Baharom[5], Sumaiya Islam[6], Sun-Ok Chung[17], Ten Sen Teik[5], Tinah Manduna Mutabazi[24], Wei-Chih Lin[9], Yeon Jin Cho[25] and Young Ho Kang[26]

1 Leibniz Institute for Agricultural Engineering and Bioeconomy (ATB), Potsdam, Germany

2 Adaptive AgroTech Consultancy, Potsdam, Germany

3 Faculty of Agriculture, Department of Land Management, University Putra Malaysia, Serdang, Selangor, Malaysia

4 Department of ICT and Natural Sciences, Norwegian University of Science and Technology (NTNU), Ålesund, Norway

5 Engineering Research Centre, Malaysian Agricultural Research and Development Institute (MARDI), Serdang, Malaysia

6 Department of Smart Agricultural Systems, Graduate School, Chungnam National University, Daejeon, Korea

7 Universiti Teknologi Malaysia, Malaysia

8 HSM Co., Ltd, Cheonan, Republic of Korea

9 Department of Mechanical and Electro-mechanical Engineering, National Sun Yat-sen University, Kaohsiung, Taiwan

10 Jeonbuk Regional Branch, Korea Electronics Technology Institute (KETI), Jeonju, Korea

11 Department of Bio-Industrial Machinery Engineering, Pusan National University, South Korea

12 National Institute of Agricultural Sciences, Rural Development Administration, Jeonju, Republic of Korea

13 FYD Company Ltd., Suwon, Republic of Korea

14 Department of Biological and Agricultural Engineering, Faculty of Engineering, Universiti Putra Malaysia (UPM), Malaysia

15 Department of Animal Science and Biotechnology, Graduate School, Chungnam National University, Daejeon, Republic of Korea

16 College of Mechanical and Electronic Engineering, Northwest A&F University, Yangling, Shaanxi, China

17 Department of Agricultural Machinery Engineering, Graduate School, Chungnam National University, Daejeon, Republic of Korea

18 Department of Agricultural and Industrial Engineering, Hajee Mohammad Danesh Science and Technology University, Dinajpur, Bangladesh

19 Geoscience and Digital Earth Centre (INSTeG), Research Institute for Sustainable Environment, UTM, Johor, Malaysia

20 Petronas Research Sdn Bhd, Malaysia

21 Chitose Institute of Science and Technology, Japan

22 University of Technology MARA, Malaysia

23 Civil Engineering Department, Indian Institute of Technology Roorkee, India

24 Innovation Village, Kampala, Uganda

25 Jeonnam Agricultural Research Extension Services, Naju, Korea

26 Jeollabukdo Agriculture Research and Extension Services, Iksan, Republic of Korea

*Address all correspondence to: rshamshiri@atb-potsdam.de

## References

[1] Martello M et al. Assessing the temporal and spatial variability of coffee plantation using RPA-based RGB imaging. Drones. 2022;**6**(10):267

[2] Binbin X, Jizhan L, Meng H, Jian W, Zhujie X. Research progress on autonomous navigation technology of agricultural robot. In: 2021 IEEE 11th Annual International Conference on CYBER Technology in Automation, Control, and Intelligent Systems (CYBER). Jiaxing, China; 2021. pp. 891-898. DOI: 10.1109/CYBER53097.2021.9588152

[3] You W et al. Data fusion of UWB and IMU based on unscented Kalman filter for indoor localization of quadrotor UAV. IEEE Access. 2020;**8**:64971-64981

[4] Liu Y, Sun R, Liu J, Fan Y, Li L, Zhang Q. Research on the positioning method of autonomous mobile robot in structure space based on UWB. In: 2019 International Conference on High Performance Big Data and Intelligent Systems (HPBD&IS). Shenzhen, China; 2019. pp. 278-282. DOI: 10.1109/HPBDIS.2019.8735462

[5] Thapa A, Horanont T. Digital twins in farming with the implementation of agricultural technologies. In: Boonpook W, Lin Z, Meksangsouy P, Wetchayont P, editors. Applied Geography and Geoinformatics for Sustainable Development. Springer Geography. Cham: Springer; 2023. DOI: 10.1007/978-3-031-16217-6_9

[6] de Queiroz DM et al. Digital Agriculture. Springer; 2022

[7] Chen C-J et al. Identification of fruit tree pests with deep learning on embedded drone to achieve accurate pesticide spraying. IEEE Access. 2021;**9**:21986-21997

[8] Kurbanov RK, Zakharova NI. Determination of spring barley lodging area with help of unmanned aerial vehicle. In: Ronzhin A, Berns K, Kostyaev A, editors. Agriculture Digitalization and Organic Production. Smart Innovation, Systems and Technologies. Vol. 245. Singapore: Springer; 2022. DOI: 10.1007/978-981-16-3349-2_21

[9] Ni X, Li C, Jiang H, Takeda F. Deep learning image segmentation and extraction of blueberry fruit traits associated with harvestability and yield. Horticulture Research. 2020;7:110. DOI: 10.1038/s41438-020-0323-3

[10] Testa R et al. Economic sustainability of Italian greenhouse cherry tomato. Sustainability. 2014;**6**(11):7967-7981

[11] Walker-Bone K, Palmer K. Musculoskeletal disorders in farmers and farm workers. Occupational Medicine. 2002;**52**(8):441-450

[12] Duncan E et al. New but for whom? Discourses of innovation in precision agriculture. Agriculture and Human Values. 2021;**38**:1181-1199

[13] Balsari P, Marucco P, Tamagnone M. A crop identification system (CIS) to optimise pesticide applications in orchards. The Journal of Horticultural Science and Biotechnology. 2009;**84**(6):113-116

[14] Raja R et al. Real-time control of high-resolution micro-jet sprayer integrated with machine vision for precision weed control. Biosystems Engineering. 2023;**228**:31-48

[15] Marras WS et al. Biomechanical risk factors for occupationally related low back disorders. Ergonomics. 1995;**38**(2):377-410

[16] Fethke NB et al. Musculoskeletal pain among Midwest farmers and associations with agricultural activities. American Journal of Industrial Medicine. 2015;**58**(3):319-330

[17] Thamsuwan O et al. Potential exoskeleton uses for reducing low back muscular activity during farm tasks. American Journal of Industrial Medicine. 2020;**63**(11):1017-1028

[18] Ji X et al. SIAT-WEXv2: A wearable exoskeleton for reducing lumbar load during lifting tasks. Complexity. 2020;**2020**:1-12

[19] Li G et al. Real-time detection of kiwifruit flower and bud simultaneously in orchard using YOLOv4 for robotic pollination. Computers and Electronics in Agriculture. 2022;**193**:106641

[20] Li G et al. Multi-class detection of kiwifruit flower and its distribution identification in orchard based on YOLOv5l and Euclidean distance. Computers and Electronics in Agriculture. 2022;**201**:107342

[21] Gao C et al. A novel pollination robot for kiwifruit flower based on preferential flowers selection and precisely target. Computers and Electronics in Agriculture. 2023;**207**:107762

[22] Zhou Z et al. Real-time kiwifruit detection in orchard using deep learning on Android™ smartphones for yield estimation. Computers and Electronics in Agriculture. 2020;**179**:105856

[23] Fu L et al. Fast and accurate detection of kiwifruit in orchard using improved YOLOv3-tiny model. Precision Agriculture. 2021;**22**:754-776

[24] Suo R et al. Improved multi-classes kiwifruit detection in orchard to avoid collisions during robotic picking. Computers and Electronics in Agriculture. 2021;**182**:106052

[25] Song Z et al. Canopy segmentation and wire reconstruction for kiwifruit robotic harvesting. Computers and Electronics in Agriculture. 2021;**181**:105933

[26] Longsheng F et al. Kiwifruit recognition at nighttime using artificial lighting based on machine vision. International Journal of Agricultural and Biological Engineering. 2015;**8**(4):52-59

[27] Wu Z et al. Coefficient of restitution of kiwifruit without external interference. Journal of Food Engineering. 2022;**327**:111060

[28] Putra AN et al. Pineapple biomass estimation using unmanned aerial vehicle in various forcing stage: Vegetation index approach from ultra-high-resolution image. Smart Agricultural Technology. 2021;**1**:100025

[29] Usman M et al. Statistical model and prediction of pineapple plant weight. Science International. 2015;**27**(2):937-943

[30] Nurul Shamimi A et al. Growth performance of selected pineapple hybrids on peat soil. Advances in Plant Science and Technology. 2017;**24**:64-67

[31] Joy P, Sindhu G. Diseases of pineapple (*Ananas comosus*): Pathogen, symptoms, infection, spread and management. In: Consultado Agosto. 2012. Available from: https://www.researchgate.net/profile/Pp-Joy/publication/306017784_DISEASES_OF_PINEAPPLE_Ananas_comosus_Pathogen_symptoms_infection_spread_management/links/57aaeb7308ae42ba52ae66cb/DISEASES-OF-PINEAPPLE-Ananas-comosus-Pathogen-symptoms-infection-spread-management.pdf

Chapter 3

# Wheat Crops Monitor: A More Reliable Rust Disease Detector for Wheat Farming

*Mohammed El Idrissi, Redmond R. Shamshiri and Ibrahim A. Hameed*


## Abstract

Crop health monitoring, as an intelligent farming activity, has become increasingly difficult and tedious for farmers due to the diversity in plant pathology. The performance of frequentist convolutional neural network (CNN) models is no longer sufficient to detect pertinent features in images and detect diseases early and efficiently. On the other hand, uncertainty in convolutional network inference is a big concern that can introduce more inexactitude in the predicted classes. In this context, we propose an intelligent farming application that aims to provide farmers with assistance in wheat crop health monitoring. The Bayesian inference with local reparameterization trick has been used to improve the sampling process during the learning phase. Thus, the uncertainty in the model and the output have been modeled to give an idea of the room for improvement. The classification skill of the proposed Bayesian uncertainty-based monitor can distinguish between wheat crops with no diseases and those infected with leaf and stem rust based on leaf and stem super-resolution image processing. The achieved accuracy is 96%, with a big resistance against overfitting/underfitting issues, and more reliability is obtained through the tolerance of the classification concept. The model is also optimized for real-time inference and adapted for resource-constrained devices.


**Keywords:** smart farming, rust disease detection, uncertainty, Bayesian inference, deep convolutional neural network

## 1. Introduction

Recently, bio-aggressors have taken diverse forms and caused significant damage to cereal production. For instance, 20% of the productivity of some sensitive wheat varieties is lost due to oidium disease, and 50% of productivity loss is caused by septoria leaf blotch disease, without forgetting pathogens that infect wheat plant's leaves and stems with different kinds of rust diseases (crown, brown, black, and yellow). Moreover, the morphological identification of diseases is unreliable, and the costs generated by human monitoring and technologies are increased. Hence, there is a necessity to have more affordable and sophisticated detection methods based on modern scientific models to accurately and early identify infections in the leaves and stems of wheat plants. Computer vision techniques in general and deep convolutional neural networks (DCNN) in a particular way are considered attractive variants that gain more attention in image processing applications and are widely used in many sophisticated applications such as face recognition [1], analyzing documents [2], understanding climate [3]. Consisting of layers and neurons interconnected with each other and signals called weights that contribute to neurons' activation, the DCNN learns from a dataset composed of linear and non-linear data samples. Noisy data, and random fluctuations in the learned data points generally cause this non-linearity. Consequently, the DCNN is susceptible to falling into overfitting or underfitting problems, which results in the model being significantly overconfident in its future decisions. Many techniques are proposed to limit the effect of overfitting-underfitting issues and introduce more generalization such as resampling or k-*fold* cross-validation technique, using the validation sub-dataset or regularization techniques such as weight decay, early stopping, L1-L2 regularization, however *dropout* function remains a widely and effective technique that is recently used to regularize models [4, 5].

### 1.1 Problem statement

Traditional neural networks perform well in most prediction and classification tasks, but they need stricter conditions to achieve these performances, such as large amounts of data and the requirement of regularization techniques. Despite that, overconfidence in decisions persists. Estimating the uncertainty in the classification tasks of the deep CNN decisions using Bayesian inference seems to be a good technique to get more confidence and trust in the decisions since it allows quantifying the degree of uncertainty in the classification tasks and introduces more regularization to the model. When applied in the smart farming domain, this feature will help farmers monitor their crops with a high level of precision. To address the problem of uncertainty in DCNNs and introduce more regularization into the network's parameters, the Bayesian approach with sampling process enhancement is used.

### 1.2 Contributions

This work capitalizes on the results of [5–7] to achieve the main following contributions:

- Introducing the Bayesian approach as a statistical and effective method of handling uncertainty.

- Investigating how random and stochastic uncertainties can be estimated.

- Applying these techniques in the smart farming field specifically to monitor the health of wheat crops with high precision throughout the tolerance of classification concept. This monitoring application is called Wheat Crops Monitor (WCM).

The rest of this paper is organized as follows: the sources of uncertainty are discussed in Section 2. The uncertainty quantification in DCNN using the Bayesian framework is developed in Section 3. The WCM model is presented in Section 4. The experiments and obtained results are discussed in Section 5. The summary and conclusions are given in Section 6.

## 2. Background

### 2.1 Uncertainty and its sources

Understanding the statistical and non-statistical uncertainty in deep artificial neural networks (ANN) is essential for many real-world applications. In statistical uncertainty, the probability distribution density follows some models that have already been proposed [8] or can be approximated in the case of intractability. While in non-statistical uncertainty, the behavior of the input data may be imprecise, which leads to an unknown probability distribution. In this case, fuzzy logic can be used to estimate the uncertainty in the prediction task [9].

The Bayesian framework offers a principle-based approach to achieve uncertainty quantification, but its application in deep ANNs is challenging due to a large number of parameters and data. Neural networks have become a very attractive discipline in so-called high-risk areas such as medical image analysis and autonomous vehicles, yet their implementation in such missions is still limited. The limitations are primarily imposed by uncertainty already included in the data (data uncertainty) or a lack of good knowledge of the deep ANN model (model uncertainty). To overcome these limitations, it is essential to estimate the uncertainty, and the uncertain predictions can then be ignored or passed on to human experts. Providing uncertainty estimates is not only important for reliable decision-making in high-risk areas but also crucial in areas where data sources are highly inconsistent and labeled data are scarce. Uncertainty estimates are also very important for areas where uncertainty is a crucial factor in the learning technique, such as active learning or reinforcement learning. In practice, these are the five factors we believe to be a vital source of uncertainty in the prediction of ANN (**Table 1**) [10].

### 2.2 Comparison with existing works

Many recent contributions have addressed methods and techniques to control plant diseases. For instance, Barbedo [11] identified key challenges in plant disease identification using digital image processing and proposed solutions to address them. Ferentinos [12] used CNN to develop a plant diseases classifier and reported a training accuracy of 99.53%. Similarly, Lu et al. [13] proposed a CNN-based method for rice disease identification, reporting a training accuracy of 95.48%. Umamageswari et al. [14] developed a PNAS (Progressive Neural Architecture Search) model to predict plant leaf disease with a training accuracy of 97.43%. Xie et al. [15] proposed an enhanced CNN to detect grape leaf diseases, achieving a training accuracy of 81.1%. Zhang et al. [16] suggested the GPDCNN model (global pooling dilated CNN) to identify plant cucumber leaf diseases, reporting an average recognition rate of 90% compared to the other tested methods. Despite the high training accuracy reported by these previous works, none of them addressed the uncertainty in the decision-making process. or declared how to deal with missing data, or how to treat the problems related to the structure of their models. The potential of this work is to detect leaf and stem rust diseases in their early stages with high reliability, making the model more robust against overfitting problems thanks to the regularization effect introduced on the model's parameters, and identify the room for improvements (in the data quality or the model structure).

| Type of uncertainty | Source of uncertainty | Signification |
|---|---|---|
| Stochastic uncertainty[1] | Errors in the specifications of the ANN architecture | The structure of an ANN has a direct effect on its performance and also on the uncertainty of its prediction. For example, the number of parameters affects the memorizing ability, which can lead to overfitting or underfitting of training data. Regarding the uncertainty in ANNs, it is known that deeper networks tend to be overly confident in their Softmax output, which means that they predict too much probability on the class with the highest probability score. |
| | Errors in the ANN learning process | The ANN learning process consists of several parameters that need to be defined (batch size, optimization function, learning rate, stopping criteria, regularization, etc.) as well as stochastic decisions in the learning process (data batch generation and weight initialization). All these decisions affect the local optimum and it is therefore very unlikely that two training processes provide the same model setting. Training data that suffer from an imbalance or low coverage of individual regions in the distribution of data also introduce uncertainties in the learned parameters. This could be mitigated by applying a data augmentation strategy to improve variety or by balancing the impact of classes or individual regions on the cost function. |
| Random uncertainty[2] | Errors caused by unknown data | Particularly in classification tasks, an ANN learned from samples from space E1 may also be able to process samples from a completely different space E2. This is the case when a network that has trained using images of bikes and cars receives a sample showing a tree. Here, the source of the uncertainty is not linked to the process of data acquisition, since it is assumed that space will contain only samples deemed to be true for a prediction task. Even if the practical result may be impacted by sensor noise or by the complete failure of these sensors, the inputs considered here represent a valid sample, but for a different task or domain. |
| | Variability in real-world applications | Most real-world environments are highly variable and almost constantly affected by changes. Changes in the environment can also affect the expression of objects (ex. plants after rain appear very different from plants after a drought). When the real-world conditions change regarding the training data, this is known as a distribution change. ANNs are sensitive to changes in distribution that can significantly impact their performance. |
| | Errors in the measurement system | The measurements themselves can be a source of uncertainty in the ANN prediction. This can be caused by insufficiency in measured information such as image resolution, or by having a false measure or insufficient information modalities. In addition, it can be caused by a noise like a sensor noise, movement, or mechanical stress, leading to inaccurate measurements. Moreover, false labeling is also a source of uncertainty that can be considered as an error and noise in the measurement system. It is referenced as a labeling noise and affects the model by reducing confidence in accurate class prediction during the training phase. |

[1]*Uncertainty in the intrinsic parameters of the model.*
[2]*Directly impacts the data collection process and introduces what is called error in observation. This type of error captures the so-called "residual or statistic" noise introduced by the physical sensors into the manipulated data, or motion noise in the case of vision systems. This type of uncertainty is divided into two families: Heteroscedastic random uncertainty (or non-equivalent variance) which assumes that for each observation (x, y) the distribution of noise is different in the data samples. On the other hand, the random Homoscedastic uncertainty (or equivalent variance) assumes that the noise is the same for each data point (x, y).*

**Table 1.**
*A summary of vital uncertainty sources.*

## 3. Deep neural networks with a probability distribution over parameters

Statistically speaking, it is recommended to consider probability distribution over the model's weights rather than assigning them specific values. This gives rise to two approaches, the classical method, where each model parameter is assigned a fixed value, resulting in a direct prediction (**Figure 1(a)**). The new Bayesian-based method (**Figure 1(b)**) computes the posterior probability distribution $wq_\theta(w|D)$ over the parameters and updates them by backpropagation. This update process uses the *Bayes-by-backdrop* framework [5, 6, 17], which is a variational inference method that allows the model to learn from the dataset the posterior probability distribution over weights and translate the estimated uncertainty into predictions. This method provides increased robustness against overfitting, a common issue in classical neural network approaches, and significantly improves model regularization. Besides this, it is very useful for applications where enough data for model training and validation are not available or available data present a lot of noise and uncertainty.

Various methods have been proposed to approximate the true posterior probability distribution in Bayesian learning. For instance, Monte-Carlo Dropout and hybrid Monte-Carlo Dropout have been utilized by Gal and Ghahramani [18]. The posterior distribution is approximated using Laplace's method [19] with a Gaussian centered at the maximum a posteriori (MAP) estimate. However, the variational inference remains the most attractive approach for approximating posterior distribution. Some attempts [20, 21] in this context have been applied using Dropout and Gaussian Dropout, respectively, in this context.

In probabilistic machine learning, the Bayesian rule is used to compute the posterior probability distribution of the model's parameters with respect to (w.r.t) θ given our dataset $p_D(\theta|x,y) = \frac{p_D(y|x,\theta)p(\theta)}{p(y|x)}$, the denominator is a normalizing constant and does not depend on θ; it is often ignored, so the posterior becomes proportional to the likelihood times the prior $p_D(\theta|x,y) = p_D(y|x,\theta)p(\theta)$. Once the posterior is computed, we are now interested in making new inferences of new outputs $y$ given new inputs, data samples $x$ using the complete Bayesian analysis described in Eq. (1) and known as the predictive distribution:

$$p_D(y|x,x,y) = \int p_D(y|x,\theta)p_D(\theta|x,y)d\theta \tag{1}$$

It was agreed that this predictive distribution is intractable due to the integration sign and the large number of parameters the model has; thus, many approximation methods have been investigated, as mentioned earlier. Bayes-by-backdrop seems to be an effective way to approximate the true posterior probability distribution with a new probability distribution and make the approximate distribution as close as possible to the true posterior. This approximation is found by minimizing the Kullback-Leibler distance between the distributions q and p w.r.t θ (Eq. (2)):

$$KL\left(q_\theta(w \vee D) \vee p(w)\right)_\theta = \int q_\theta(w \vee D)\, log_2 \frac{q_\theta(w|D)}{p(w)} dw \tag{2}$$

Eq. (2) forms another optimization problem that must be minimized w.r.t θ, with the resulting cost function known as variational free energy [22] and given by Eq. (3):

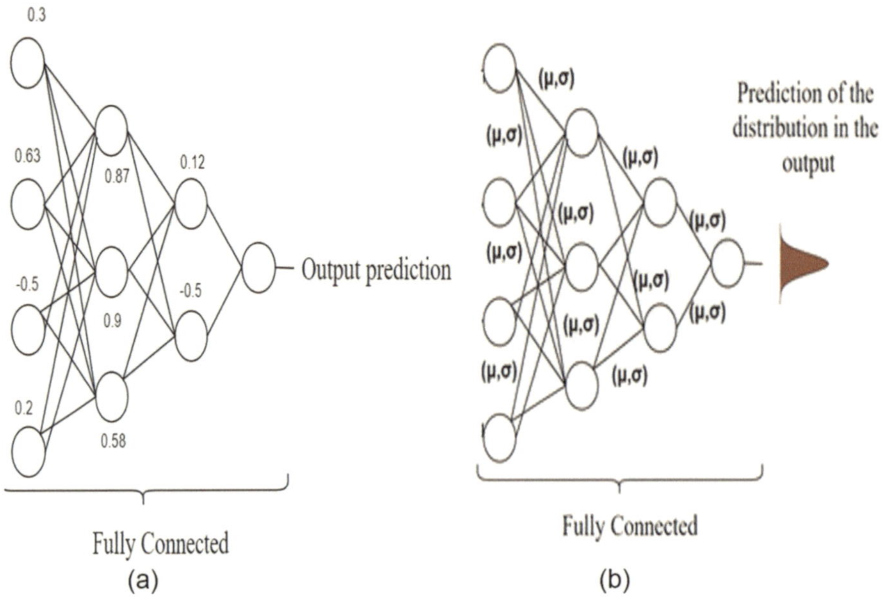


**Figure 1.**
*Frequentist method; (a) versus Bayesian approach; and (b) in parameters tuning (the numbers in (a) are indicative).*

$$\theta_{opt} = argminKL\left(q_\theta(w \vee D) \vee p(w \vee D)\right)_\theta$$
$$argminKL\left(q_\theta(w \vee D) \vee p(w)\right)_\theta - E_{q(w|D)}[logp(D|w)] + logp(D) \quad (3)$$

The variational free energy is built upon three terms: the first term is called complexity cost, the second term is known as likelihood cost, and the last term is a constant that is ignored during the learning process [23]. This cost function is intractable, and it is unable to be computed exactly. A stochastic variational method [17] is adopted to arrive finally at this tractable form of the objective function given by Eq. (4) [5, 24]. This equation should be optimized during the Bayesian learning process.

$$F(D, \theta) \approx \sum_{i=1}^{N} log\, q_\theta\left(\omega^{(i)}|D\right) - log\, p\left(\omega^{(i)}\right) - log\, p\left(D|\omega^{(i)}\right) \quad (4)$$

Where n represents the number of samples $\omega^{(i)}$ drawn from the approximated variational distribution. The first term of this equation is known as the variational posterior and is taken as the log of the Gaussian distribution with mean μ and variance $\sigma^2$, so $log\,(q|\theta(w \vee D)) \sum log\,(N(w|\mu, \sigma^2))$. The second term is the prior and is taken as the log of many elementary Gaussians, so $log\,(p(w)) \sum log\,(N(w|0, \sigma^2))$. The last term is a vector of likelihood probabilities that is computed using the Softmax function. The final statistical and stochastic uncertainties are expressed by Eq. (5) [25, 26]:

$$U = \frac{1}{N}\sum_{n=1}^{N} diag(\hat{p}_n) - \hat{p}_n\hat{p}_n^N + \frac{1}{N}\sum_{n=1}^{N}(\hat{p}_n - \bar{p})(\hat{p}_n - \bar{p})^N \quad (5)$$

Where $\hat{p}_n$ is a vector of probabilities computed using the Softmax function and $\bar{p}$ is the mean value of $\hat{p}_n$. The sampling process is improved using the local reparameterization trick (LRT) [5], which is a commonly used method in statistical problem rewriting [7]. This technique directly samples parameters and consumes more time and resources. Instead, we sample from the approximate variational distribution to win in terms of time and resources required for processing.

## 4. Application in smart farming through the wheat crops monitor (WCM)

Before describing the WCM, we remember an important point of Jayesh Bapu Ahire [27] says that the common architecture of classical deep convolutional neural network stacks a series of Convolution-Activation (Conv-Act) layers, followed by pooling layers. This series is repeated till having an image with a minimum of features, then sends them to the fully connected (FC) layers; the last FC layer holds predicted outputs. Famous deep CNN models are based on the following form:

$$Input \Rightarrow [[Conv - Act] * \mathrm{M} \Rightarrow Pooling] * \mathrm{N} \Rightarrow [\mathrm{FC} - Act] * \mathrm{L} \Rightarrow \mathrm{FC}$$

Where M, N, and L are the times each bloc of layers is repeated in the deep CNN model. Smart wheat crops health monitor schema also follows this architecture, with the particularity of considering weights, biases, and kernels as random variables and computing the probability distribution over them instead of working with single-point estimate values. The following **Figure 2** represents the structure of the model:

In this schema, we work only with probability distributions. Instead of computing a direct and single convolution operation between a small portion of the image and the kernel, we perform two convolution operations. In the first

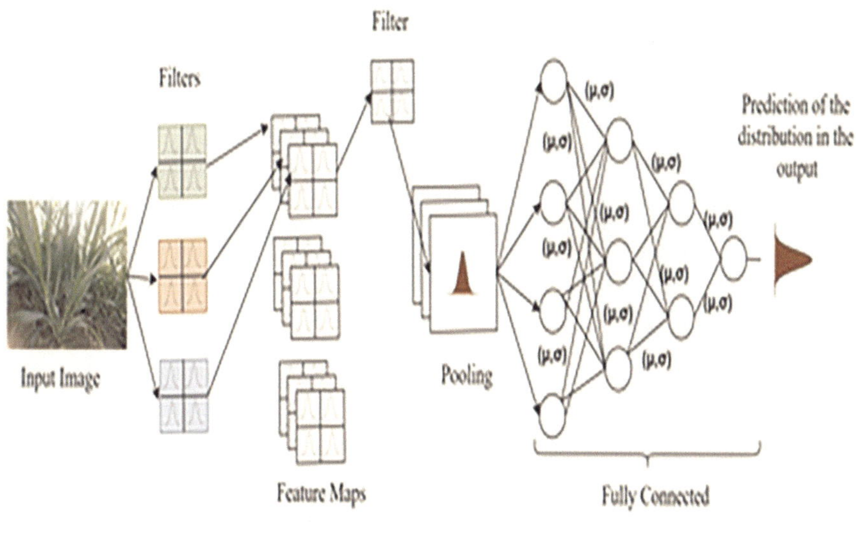


**Figure 2.**
*Typical DCNN structure showing the distribution (mean, variance) of the weights instead of single values.*

convolution, we calculate the mean $\boldsymbol{\mu}$ of the variational posterior probability distribution considered for each parameter, and this means that we compute the maximum a posteriori probability (MAP) around the parameter [5]. In the second convolution operation, the variance $\boldsymbol{\sigma} = \boldsymbol{\alpha}.\boldsymbol{\mu}^2$ is learned, this quantity consists of the pre-calculated mean and the variable α that is also learned by the convolution operation of the weights [28], and the Bayes by backpropagation algorithm is used to compute the posterior probability distributions. The local reparameterization trick assumes that for a fixed input and a Gaussian distribution over fully connected layers' weights, the obtained distribution of activations (the sampling is directly performed over activation distributions instead of calculated individually for each weight) is also a Gaussian distribution [7]. The success of this structure requires that the weights in the fully connected layers need to have probability distributions as well.

To be more precise, we decided to introduce a new concept called tolerance of classification, which aims to express how confident the true predicted labels are. The classification decision is expressed relative to a certain level of uncertainty. So, the classification result has always this form:

$$Class = predictedlabel \pm Uncertainty \tag{6}$$

In such a way, we can have an idea about how far or near our classification is from the most precise decision. The closer the estimated uncertainty can be to zero, the better the decision is.

## 5. Experiment and discussion

### 5.1 Parameters initialization

As mentioned earlier, we should approximate the true posterior probability distribution. For that, we consider the Gaussian process as an approximate variational probability distribution, and we use it as prior over the weights to compute the mean and the variance of each parameter of the model. Then, the gradient descent is computed over the random variables θ and over each weight individually. The Bayes-by-backprop is used to sample the weights and update the approximated variational posterior by backpropagation, and the LRT is applied to improve the sampling process. Adam optimizer is used in all phases of the experiment, and the Softplus function is used as an activation function since it represents the smooth approximation of the ReLu activation function. The rest of the Bayesian initial parameters are summarized in **Table 2** in Section 5.3.

### 5.2 Model compression

The approach used in this work goes back to estimating two parameters $(\mu, \sigma)$, which means that the overall number of parameters in the model will be doubled; consequently, the model will become heavier in processing. Model pruning is a concept that was proposed to introduce more optimization into the model by reducing the number of parameters and thus decreasing the complexity and resources required for processing. Here, we used the L1 normalization technique, and this method consisted of two actions [5]:

| Parameter | Value | Range of possible values |
|---|---|---|
| Number of epochs | n_epochs = 300 | n_epochs ∈ R |
| Activation function | Softplus | {Softplus, Relu} |
| Model layer type | Local Reparametrization Trick | {Bayes by Backprop (BBB), Local Reparametrization Trick (BBB_LRT} |
| Initial configuration of the priors | Initial μ = 0.01 | μ should be small |
| | Initial σ = 0.2 | σ should be small |
| Initial configuration of the posterior | Initial μ (mean, standard deviation) = (0.01, 0.01). | The couple (mean, standard deviation) should have small values |
| | Initial ρ (mean, standard deviation) = (−3, 0.01) | The couple (mean, standard deviation) should have small values |
| Initial learning rate | 10E-3 | lr_initial should be small |
| Number of workers | 5 | num_workers ∈ R |
| Batch size | 32 | batch_size ∈ R |
| Train ensemble | 1 | train_ens ∈ R |
| Validation ensemble | 1 | valid_ens ∈ R |
| Beta type | 0.01 | Beta_type can be float or ∈ {Blundell, Soenderby, Standard} |
| Dataset | Size = 11,420, Train = 57%, Validation = 38%, Test = 5% | |

**Table 2.**
*A summary of initial hyperparameters used in training and testing of the WCM model.*

- In the first action, we halved the number of filters supposed to be used in the convolution layers without impacting the model's overall performance.

- The second point consists of defining a threshold and only keeping the weights above this threshold. The other weights under the threshold are eliminated. The defined threshold and the technique of halving the number of filters are still points that need to be empirically argued. However, experimentally, it allows model pruning without impacting the overall performance of the model.

## 5.3 The material used in this work

### *5.3.1 Hardware and software*

The WCM algorithm is coded in the Python programming language using the Torch library. Torch is a high-level framework entirely written in Python. It can highly interact with neural network and machine learning libraries and provides two high-level components: tensor calculation and deep neural network manipulation, as well as it is compatible with CPU and GPU (Graphics Processing Unit) processor architectures. WCM was trained on a UCS (Unified Computing System) cluster. Each node of the cluster has an Intel(R) Xeon(R) CPU E5–2630 v4 @ 2.20GHz and 256 gigabytes of RAM, and this configuration is suitable to finish each training epoch in less than 5 minutes, instead of 15 minutes for usual computers.

### *5.3.2 Dataset*

The model provided by [5] is tested in small and large datasets: MNIST, CIFAR10, and CIFAR100 using models: AlexNet, LeNet, and 3Conv3FC. Since we work with a noMNIST dataset, many hyperparameters are reconfigured notably activation function type, layer type, and the number of batches for which uncertainty needs to be estimated. The model is trained and tested using the CGIAR Computer Vision for Crop Disease [29] dataset. The images were collected from different sources, and a great part was gathered by CIMMYT and CIMMYT partners in Ethiopia and Tanzania. The remainder of the data is sourced from public images found on Google Images. The training, validation, and testing sub-datasets consecutively represent 57%, 38%, and 5% of the overall size of the dataset.

## 5.4 Evaluation and validation of the model

As mentioned previously, the implementation provided by [5, 30] was tested using MNIST and CIFAR datasets and using AlexNet, LeNet, and 3Conv3FC models. To build our model, we inherited the *ModuleWrapper* module, and we used the *BBBLinear* and *BBBConv2d* functions in all the BayesianCNN layers. The number of layers in the customized model is increased as well from three convolutions, three fully connected to four convolutions, and four fully connected. The super-resolution technique [31] with the Bayesian approach [5] is used to recover high resolution from low-resolution images to extract more pertinent features in the images. The configurations of the initial Bayesian network's hyperparameters are summarized in **Table 2**.

At the end of the training process, we get the accuracy and the loss function curves of the training and validation processes, as shown in **Figures 3** and **4**. We observe from the accuracy function curve (**Figure 3**) that the accuracy value of the training process begins from 77.5% and continues increasing till the third epoch, where it goes back to decreasing. From this step, we note many fluctuations between 85% and 88.75%. The maximal accuracy value is recorded in the 170th epoch; beyond this phase, accuracy begins decreasing, and this is a sign of model convergence. Regarding the cost function (**Figure 4**), the training loss value starts from 34,464,858 and continues to decrease exponentially; the minimum value is observed in the 200th

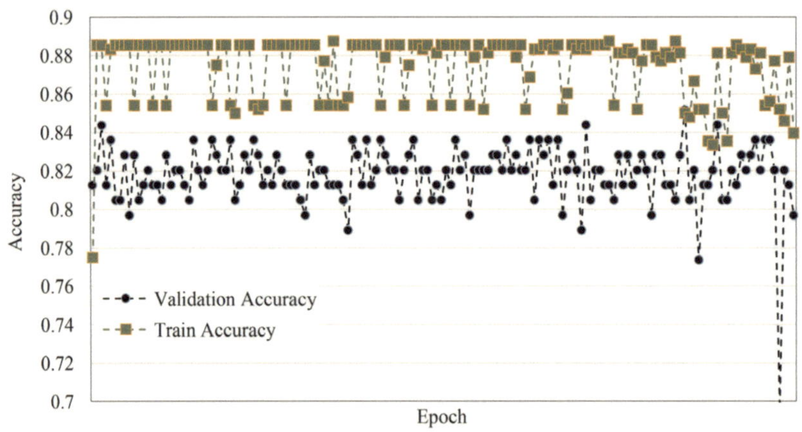


**Figure 3.**
*Training and validation accuracy functions.*

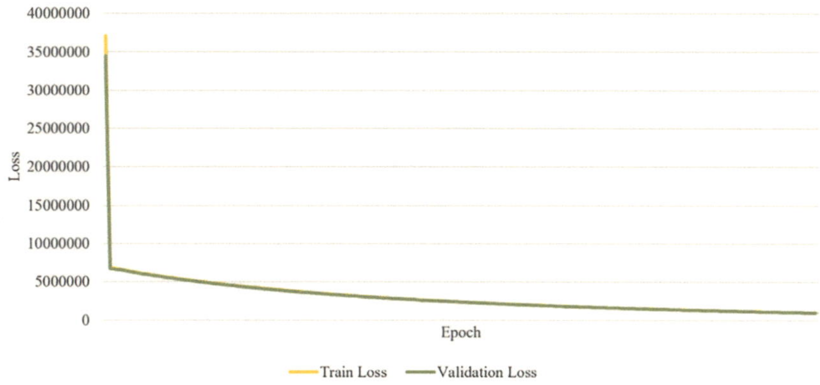


**Figure 4.**
*Training and validation loss functions.*

epoch, where it achieves 1,016,792. The curves seem to be flattened from the 170th epoch, and this is the exact point that coincides with the convergence of the model.

### *5.4.1 Metrics for evaluation*

**Table 3** presents the metrics used for the evaluation of our model. Since the WCM detects the leaf and stem rust diseases in the wheat plants and provides more reliability of detection by estimating the uncertainty, the precision, sensitivity, and F1-measure values obtained in the test phase consecutively are 96%, 92.35%, and 93.84%. Each metric is obtained by calculating the average of all classes from the confusion matrix shown in **Figure 5**. The obtained values are good and reflect the effect of the uncertainty quantification on the classification task using deep CNN. The sensitivity is significantly good because the model has shown good skills in identifying the infected samples. Some difficulties can be encountered in differentiating between leaf rust and stem rust, but generally, the performance of detecting diseases is high. **Figure 6** shows the receiver operating characteristic (ROC) of the proposed model. The global area under the curve-ROC (AUC-ROC) value of all classes is 96%, which proves the high separability performance. The individual AUC values of healthy wheat, leaf rust, and stem rust classes successively are 98.77%, 93.58%, and 95.36%.

| Metric | Designation | Mathematical formula | Variables |
|---|---|---|---|
| Precision (P) | The ratio of infected wheat plants correctly detected among all wheat plants detected as infected. | $P = \frac{TI}{TI+FI}$ (8) | • TI: represents the number of true infected wheat plants classified as infected.<br>• FI: denotes the incorrectly predicted samples as infected.<br>• FH: represents the incorrectly predicted samples as healthy. |
| Sensitivity (S) | The ratio of infected wheat plants correctly detected among all true infected wheat plants. | $S = \frac{TI}{TI+FH}$ (9) | |
| F1-Measure (F1) | The balanced mean between precision and sensitivity. | $F1 = 2\left(\frac{P.S}{P+S}\right)$ (10) | |

**Table 3.**
*Metrics used for model evaluation.*

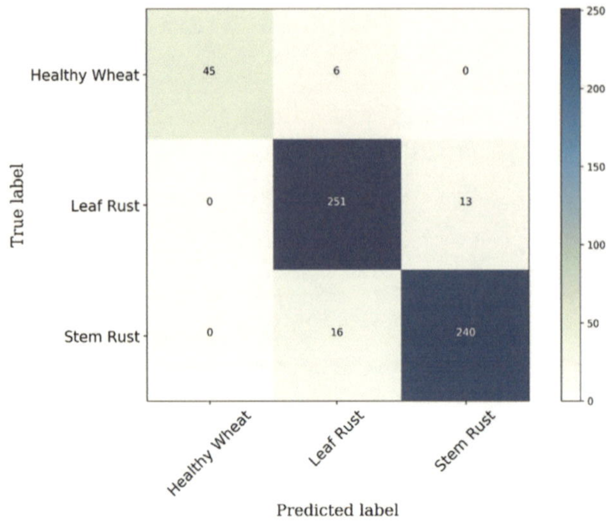


**Figure 5.**
*Confusion Matrix of WCM with uncertainty quantification.*

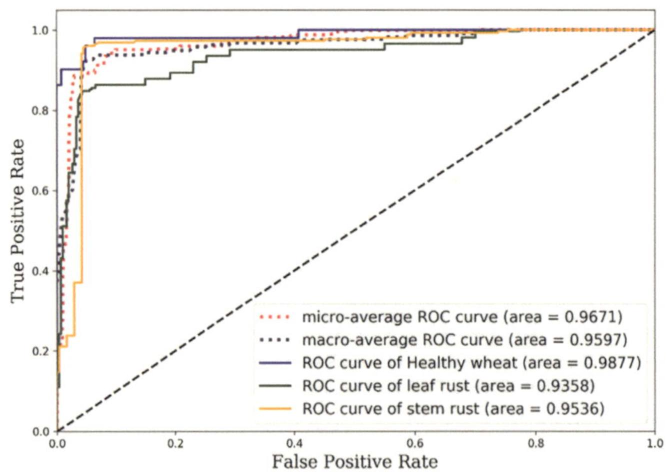


**Figure 6.**
*Receiver operating characteristic of WCM with uncertainty estimates.*

To test the performances of the uncertainty estimation of the proposed model, we have prepared **Table 4** of 8 new samples that the model has never seen before. These new samples are streamed from the test repository, six of which are healthy wheat plants, and two of which are infected with leaf and stem rust (red rust) diseases. We first prepared the new samples by making the required transforms including the high-resolution recovery, and then injected them into the input of the pre-trained model. Uncertainties and probabilities are handled using a temporary text file and compared to the inference made by a classical frequentist model prepared for this purpose. The results shown in **Table 4** prove the great classification skills (test accuracy in T6 could achieve 95.74% in variational inference) with a big precision provided by the total estimated uncertainty (statistical accumulated with stochastic in the same test T6 is estimated to be only 19%).

| **Metrics for evaluation:** | | | | | | |
|---|---|---|---|---|---|---|
| Test Samples, [Left: T1, Middle: T2, Right: T3] |  | |  | |  | |
| Inference | Variational | Frequentist | Variational | Frequentist | Variational | Frequentist |
| Accuracy of prediction | 77.75% | 80.25% | 85.21% | 87% | 62.37% | 61.98% |
| True class | Healthy | | Healthy | | Infected with leaf rust | |
| Predicted class | Healthy | Healthy | Healthy | Healthy | Infected with leaf rust | Infected with leaf rust |
| Statistical | 0.18579143 | Not applicable | 0.17867781 | Not applicable | 0.18753554 | Not applicable |
| Stochastic | 0.016439945 | Not applicable | 0.017931558 | Not applicable | 0.013076332 | Not applicable |
| Total uncertainty | 0.202231375 | Not applicable | 0.196609368 | Not applicable | 0.200611872 | Not applicable |
| Test Samples [Left: T4, Middle: T5, Right: T6] |  Garg [32] | |  | |  | |
| Inference | Variational | Frequentist | Variational | Frequentist | Variational | Frequentist |
| Accuracy of prediction | 76.16% | 70.44% | 37.27% | 66.01% | 95.74% | 92.45% |
| True class | Infected with stem rust | | Healthy | | Healthy | |
| Predicted class | Infected with stem rust | Infected with stem rust | Infected with leaf rust | Healthy | Healthy | Healthy |
| Statistical | 0.18301016 | Not applicable | 0.1900446 | Not applicable | 0.18442893 | Not applicable |
| Stochastic | 0.010087308 | Not applicable | 0.011339135 | Not applicable | 0.009034724 | Not applicable |
| Total uncertainty | 0.19309747 | Not applicable | 0.201383739 | Not applicable | 0.193463654 | Not applicable |
| Test Samples [Left: T7, Right: T8] |  | | |  Yara UK [33] | | |

| Inference | Variational | Frequentist | Variational | Frequentist |
|---|---|---|---|---|
| Accuracy of prediction | 71.91% | 70.22% | 87.90% | 80.15% |
| True class | Healthy | | Healthy | |
| Predicted class | Healthy | Healthy | Healthy | Healthy |
| Statistical | 0.18126002 | Not applicable | 0.19115148 | Not applicable |
| Stochastic | 0.009895712 | Not applicable | 0.016715055 | Not applicable |
| Total uncertainty | 0.191155731 | Not applicable | 0.2078665346 | Not applicable |

**Table 4.**
*Test of the prediction and uncertainty estimation function compared to the classical frequentist inference.*

## 5.5 Discussion

In our adopted approach, we assumed that the weights of the DCNN have a probability distribution instead of fixed values. Due to the great difficulty of applying the Bayesian framework to infer the Bayesian distribution on weights, it is often necessary to use the Bayes-by-backprop algorithm. The use of this technique to extract the variational approximation of the true posterior distribution is not only a method to make our model much more accurate in quantifying global uncertainty, but it also leads to the regularization of the model without having to use the dropout function. The utilization of this method to measure the uncertainty caused by the model or by data helps to improve the reliability of the obtained results, which is why the output of each classification decision is expressed using Eq. (6). **Table 4** summarizes an example of tests coded $T_i$ that we performed using images collected from the test field, and others downloaded from the internet. To better explain the results, let us take the test code T6. The output, according to Eq. (6), is 95.74% $\pm$ 19.35% healthy against 92.45% for the frequentist inference. This decision is composed of two terms: the accuracy of classification and uncertainty about this accuracy. Thus, the image is seen as healthy, with an accuracy of 95.74% and a doubt of 19.35%. The details of the estimated uncertainty are given by the statistical and stochastic values. Another significant example is test T5, which is a wrong classification decision. In this scenario. The model classified the image as infected with leaf rust with a probability of 37.27% and a doubt level equal to 20.14% compared to 66.01% healthy for frequentist inference. However, the truth is that this is a healthy wheat plant that suffers from vegetation issues due to the drought, causing the distribution of plants to be non-uniform and appear infected. This is one of the factors that can lead to false positives. In reality, these false positives highlight issues other than diseases. These false positives highlight issues other than diseases, such as drought or nutrient deficiencies, which require the attention of the farmer or the field manager.

# 6. Summary conclusions

In this study, we introduce a novel promising wheat crops monitoring system called WCM, employing the Bayesian framework built upon a customized deep CNN algorithm. The technique utilized in this research, referred to as Bayes by backpropagation and Gaussian distribution, embodies Bayesian characteristics. Initially, samples undergo transformation to better suit the model's requirements, generating high-quality images from low-resolution ones. Subsequently, super-resolution images are integrated into the model's input for training and validation. The achieved validation accuracy is 88.75%, whereas the test accuracy could achieve up to 95% with an uncertainty margin precisely estimated. The reliability of the system is proved by its ability to provide the predicted class and the level of doubt in the classification task, as well as to direct us toward the improvement of the performance through uncertainty estimation. The real implementation of this monitor can be done using a drone to visually scan the field, then send the captured real-time images to the WCM to draw the wheat global health map and send results back to the farmer, who can make early interventions to make pesticides in early stages and in reasonable quantities to save the wheat crops.

## Author details

Mohammed El Idrissi[1]*, Redmond R. Shamshiri[2] and Ibrahim A. Hameed[3]

1 Computer Science, Informatics, Signals, Automatics, and Cognitivism Laboratory, Sidi Mohamed Ben Abdellah University Fez-Atlas, Morocco

2 Leibniz Institute for Agricultural Engineering and Bioeconomy, Potsdam, Germany

3 Department of ICT and Natural Sciences, Norwegian University of Science and Technology (NTNU), Ålesund, Norway

*Address all correspondence to: mohammed.elidrissi29@usmba.ac.ma

## References

[1] Li HC, Deng ZY, Chiang HH. Lightweight and resource-constrained learning network for face recognition with performance optimization. Sensors. 2020;**20**:6114. DOI: 10.3390/s20216114

[2] Borges Oliveira DA, Viana MP. Fast CNN-based document layout analysis. Proceedings IEEE International Conference on Computer Vision Workshops, Venice, Italy. 2017; **1173-1180**. DOI: 10.1109/ICCVW.2017.142

[3] Chattopadhyay A, Hassanzadeh P, Pasha S. Predicting clustered weather patterns: A test case for applications of convolutional neural networks to spatio-temporal climate data. Scientific Reports. 2020;**10**(3):1317. DOI: 10.1038/s41598-020-57897-9

[4] Hinton GE, Srivastava N, Krizhevsky A, Sutskever I, Salakhutdinov RR. Improving neural networks by preventing co-adaptation of feature detectors. arXiv preprint arXiv:1207.0580. 2012

[5] Kumar S, Felix L, Marcus L. A Comprehensive Guide to Bayesian Convolutional Neural Network with Variational Inference. arXiv preprint. arXiv:1901.02731. 2019

[6] Blundell C, Cornebise J, Koray Kavukcuoglu K, and Wierstra D. Weight Uncertainty in Neural Networks. arXiv preprint arXiv:1505.05424. 2015

[7] Kingma DP, Salimans T, Welling T. Variational Dropout and the Local Reparameterization Trick. arXiv:1506.02557. 2015

[8] Ronald HD. Measurement uncertainty models. ISA Transactions. 1997;**36**:29-35. DOI: 10.1016/S0019-0578(97)00004-9

[9] Xia X, Wang Z, Yongsheng G. Estimation of non-statistical uncertainty using fuzzy-set theory. Measurement Science and Technology. 2000;**11**: 430-435. DOI: 10.1088/0957-0233/11/4/314

[10] Gawlikowski J, Tassi CRN, Ali M, Lee J, Humt M, Feng J, Kruspe A, Triebel R, Jung P, Roscher R, Shahzad M, Yang W, Bamler R, Zhu XX. A Survey of Uncertainty in Deep Neural Networks. arXiv:2107.03342. 2021

[11] Barbedo JGA. A review on the main challenges in automatic plant disease identification based on visible range images. Biosystems Engineering. 2016; **144**:52-60. DOI: 10.1016/j.biosystemseng.2016.01.017

[12] Ferentinos KP. Deep Learning Models for Plant Disease Detection and Diagnosis. Computers and Electronics in Agriculture. 2018;**145**:311-318. DOI: 10.1016/j.compag.2018.01.009

[13] Lu Y, Yi S, Zeng N, Liu Y, Zhang Y. Identification of rice diseases using deep convolutional neural networks. Neurocomputing. 2017;**267**:378-384. DOI: 10.1016/j.neucom.2017.06.023

[14] Umamageswari A, Bharathiraja N, Shiny Irene D. A novel fuzzy C-means based chameleon swarm algorithm for segmentation and progressive neural architecture search for plant disease classification. ICT Express. 2021;**9**: 160-167. DOI: 10.1016/j.icte.2021.08.019

[15] Xiaoyue X, Yuan M, Bin L, Jinrong H, Shuqin L, Hongyan W. A deep-learning-based real-time detector for grape leaf diseases using improved

convolutional neural networks. Frontiers in Plant Science. 2020;**11**:751. DOI: 10.3389/fpls.2020.00751

[16] Shanwen Z, Subing Z, Chuanlei Z, Xianfeng W, Yun S. Cucumber leaf disease identification with global pooling dilated convolutional neural network. Computers and Electronics in Agriculture. 2019;**162**:422-430. DOI: 10.1016/j.compag.2019.03.012

[17] Graves A. Practical variational inference for neural networks. In: Shawe-Taylor J, Zemel R, Bartlett P, Pereira F, Weinberger KQ, editors. Curran Associates Inc: Advances in Neural Information Processing Systems; 2011. pp. 2348-2356

[18] Gal Y, Ghahramani Z. Bayesian Convolutional Neural Networks with Bernoulli Approximate Variational Inference. arXiv preprint arXiv: 1506.02158. 2015

[19] Friston K, Ashburner J, Kiebel S, Nichols T, Penny W. Statistical Parametric Mapping: The Analysis of Functional Brain Images. Academic Press; 2006

[20] Srivastava N, Hinton G, Krizhevsky A, Sutskever I, Salakhutdinov R. Dropout: A simple way to prevent neural networks from overfitting. The Journal of Machine Learning Research. 2014;**15**:1929-1958

[21] Wang S, Manning C. Fast dropout training. In: Dasgupta S, McAllester D, editors. Proceedings of the 30th International Conference on Machine Learning. Atlanta, Georgia: PMLR; 2013. pp. 118-126

[22] Shockley JM, Dillon CR, Stombaugh T. A whole farm analysis of the influence of auto-steer navigation on net returns, risk and production practices. Journal of Agricultural and Applied Economics. 2011;**43**:57-75. DOI: 10.1017/S1074070800004053

[23] Shridhar K, Laumann F, Liwicki, M. Uncertainty Estimations by Softplus Normalization in Bayesian Convolutional Neural Networks with Variational Inference. arXiv:1806.05978. 2018

[24] Gluon MXnet. Chapter18 Variational-Methods-and-Uncertainty. 2022. Available from: https://gluon.mxnet.io/chapter18_variational-methods-and-uncertainty/bayes-by-backprop.html [Accessed: 26 January 2023]

[25] Kendall A, Gal Y. What uncertainties do we need in bayesian deep learning for computer vision? Advances in Neural Information Processing Systems. 2017: 5574-5584

[26] Kwon Y, Won JH, Kim BJ, Paik MC. Uncertainty quantification using bayesian neural networks in classification: Application to ischemic stroke lesion segmentation. Medical Imaging with Deep Learning; 2018

[27] Ahire JB. Artificial Neural Networks: The Brain Behind AI. Available from: Lulu.com; 2018

[28] Molchanov D, Ashukha A, Vetrov D. Variational Dropout Sparsifies Deep Neural Networks. arXiv preprint arXiv: 1701.05369. 2017

[29] CGIAR Computer Vision for Crop Disease, 2020. Available from: https://www.kaggle.com/shadabhussain/cgiar-computer-vision-for-crop-disease/ [Accessed: 20 May 2021]

[30] GitHub – kumar-shridhar/PyTorch-BayesianCNN, 2020. Bayesian Convolutional Neural Network with

Variational Inference based on Bayes by Backprop in PyTorch. Available from: https://github.com/kumar-shridhar/PTorch-BayesianCNN/ [Accessed: 25 May 2021]

[31] Shi W, Caballero J, Huszar F, Totz J, Aitken AP, Bishop R, Rueckert D, Wang Z. Real-Time Single Image and Video Super-Resolution using an Efficient Sub-Pixel Convolutional Neural Network. CoRR, abs/1609.05158. 2016

[32] Garg S. Stem Rust – Alchetron, The Free Social Encyclopedia. 2021. Available from: https://alchetron.com/Stem-rust#stem-rust-d0c8800c-4178-4ebe-a94e-ed4e3a6e8a3-resize-750.jpeg [Accessed: 14 September 2021]

[33] Yara UK. How to Maintain Wheat Health with Nutrition|Yara UK. 2021. Available from: https://www.yara.co.uk/crop-nutrition/wheat/maintaining-wheat-health/ [Accessed: 20 September 2022]

Chapter 4

# Use of RGB Images in Field Conditions to Evaluate the Quality of Pastures in Farms in Antioquia: A Methodology

*Valentina Parilli-Ocampo, Manuela Ortega Monsalve, Mario Cerón-Muñoz, Luis Galeano-Vasco and Marisol Medina-Sierra*

## Abstract

The use of RGB (Red, Green, and Blue) images is a useful technique considered in the prediction of diseases, moisture content, height, and nutritional composition of different crops of productive interest. It is important to adopt a methodology in the field that allows the acquisition of images without losing the quality of the information in the RGB bands since the prediction and adjustment of the grass quality parameters depend on it. Currently, there are few studies and methodologies that support the validity of the use of RGB images in the field, since there are many environmental factors that can distort the information collected. For this study, a field methodology was established where RGB images were captured using the unmanned aerial vehicle drone, DJI Phantom 4 Pro. A total of 270 images of grass crops for animal feed were taken on 15 farms in Antioquia. The images were pre-processed using the programming language Python, where a region of interest for each image was chosen and the average RGB values were extracted. Different indices were created with the RGB bands and based on them; several models were used for the nutritional variables of the pasture, managing to find suitable equations for acid detergent fiber, crude protein, and moisture.

**Keywords:** drone, region of interest, spectral bands, statistical models, vegetal nutrition, weather

## 1. Introduction

Light is characterized as visually assessed radiant energy exhibiting diverse wavelengths, perceived as different colors, and emitting electromagnetic radiation that is relatively balanced in all the wavelengths contained in the visible region [1].

The color of an object is determined by the selective absorption of different amounts of wavelengths within the visible spectrum. As indicated in **Figure 1**, the

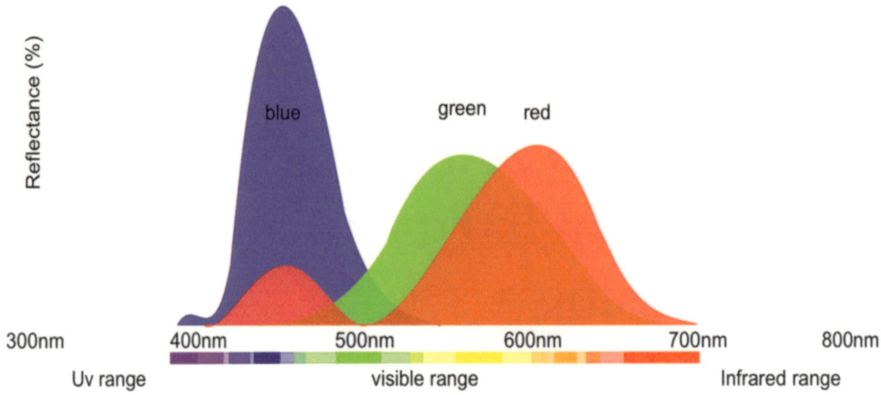


**Figure 1.**
*Representation of spectrophotometric curves for red, green, and blue wavelengths (adapted from Eissa et al. [1] and Pharr et al. [2]).*

blue, green, and red colors of an object are quantified and depicted through spectrophotometric curves. These curves serve as graphical representations of the fractions of incident light, whether reflected or transmitted, as a function of wavelength across the visible spectrum [1, 2].

Sensors using RGB cameras have evolved into widely used tools in the fields of artificial vision and agriculture. These work by means of passive cameras that are transported by an unmanned aerial vehicle (UAV) and where natural light is used to take the images [3]. Advances in digital agriculture have made it possible to extract information about the "vegetation indexes" (VI) of the crops, which are mathematical operations based on spectral data of the radiation emitted by the vegetation at various wavelengths [4]. These operations provide an understanding of characteristics such as the height of grasses by simply flying over the terrain and obtaining a photograph of the site.

Experts in the agricultural industry have used unmanned aerial vehicles (UAVs), such as drones, to analyze pixels and RGB bands to identify nutritional factors in crops. These types of techniques bring ease of use, tracking areas that are difficult for humans to access, and monitoring different crop yields [5]. Studies have shown that UAVs can be used for monitoring cultivation, with the advantage that the light source used is natural [6]. The use of UAVs has reached high levels of reliability, mapping areas of different sizes, becoming a solution and alternative at the farm level [7]. However, not enough research studies have been carried out showing a clear and applicable methodology in the field along with adequate processing to verify the accuracy of the data collected in the field. The use of RGB images in the field is intended to replace the bromatological analysis of the plants, which play a very important role in determining the nutritional composition of the plants before animals feed from them; moreover, there is a relationship between the elemental content of the plant and its nutritional status [8], knowing and quantifying the nutrient content of pastures is fundamental to decision making and establishing adequate crop management in agricultural productions [9]. Bromatological analysis performed on pastures is commonly used to determine the content of protein, fiber, sugars, and lipids, among others. However, the RGB imaging approach additionally analyses different characteristics through the image bands, allowing us to know, for example, the nutritional status of a plant [10].

The high spatial resolution and low cost of RGB cameras give them a great advantage. In general, RGB cameras have been used extensively in pasture studies with UAVs [11]. When taking images in the field, it's important to consider several factors such as the weather, light intensity, light angle, and altitude of the UAVs. According to authors like [12], these factors may have negative effects, making it challenging to conduct flights in unpredictable field conditions such as lighting, shooting angle, and altitude. The flight methodology proposed by Tagle Casapía et al. [13] suggests that the aircraft should be unmanned, and the pilot should be aware of the aircraft's altitude, as well as any previously marked locations in the field. Additionally, a calibration reflectance panel was used before and after the flight, as this data is essential to the construction of the model [14]. Overall, it is important to regulate certain field variables to obtain high-quality images and data, while maintaining a functional methodology that can be used by other researchers.

In addition to capturing images in the field, artificial intelligence (AI), intelligent automated systems (IAS) and machine learning models are necessary tools that help extract the information captured. Through the analysis of parameters, these tools also have the potential to help improve crops. They can streamline operations, observe environmental conditions, and predict the quantity and quality of yields [15, 16]. This chapter's goal was to provide a system for using RGB images of the field to assess the quality of pastures on farms in Antioquia.

## 2. Methodology for the acquisition of RGB images in the field

Below is a practical example of applying the methodology for the collection of field images using unmanned aerial vehicles which can be replicated. A series of steps and suggestions are provided that are considered necessary to successfully complete the process. This is followed by a practical example of statistical processing of RGB data to fit several models.

### 2.1 Study area

The flights and bromatological analyses of the grass were carried out in different municipalities of the Department of Antioquia. The farms selected for this study, the number of bromatological analyses carried out, and the number of photographs taken with the drone in each of them are shown in **Table 1**. The only criteria considered in choosing the farms were their accessibility and the presence of pastures for dairy and beef animals.

### 2.2 Flight planning

Planning a drone flight involves several steps that are critical to its success (**Figure 2**). These include the selection of the responsible crew, the appropriate choice of aircraft and flight programs, and the precise definition of the area of interest being covered. Each of these steps plays a crucial role in ensuring the efficient and effective execution of the flight to achieve optimal results.

#### *2.2.1 Crew selection*

In the use of a UAV, key roles are defined that form essential functions to ensure a safe and efficient operation. The pilot supervises operations and plans flight routes,

| Subregion | Municipality | Farms | Bromatological analysis |
|---|---|---|---|
| West | Sopetrán | 2 | 2 |
| Magdalena Medio | Maceo | 1 | 2 |
| East | Abejorral | 2 | 4 |
| North | Yarumal | 2 | 4 |
| Southwest | Amagá | 2 | 4 |
| North | Santa Rosa de Osos | 2 | 4 |
| Urabá | Chigorodó | 2 | 4 |
| Bajo Cauca | Caucasia | 2 | 4 |
| Southwest | Támesis | 1 | 2 |

**Table 1.**
*Region and municipality, sample size, and number of images for the properties chosen for the current study.*

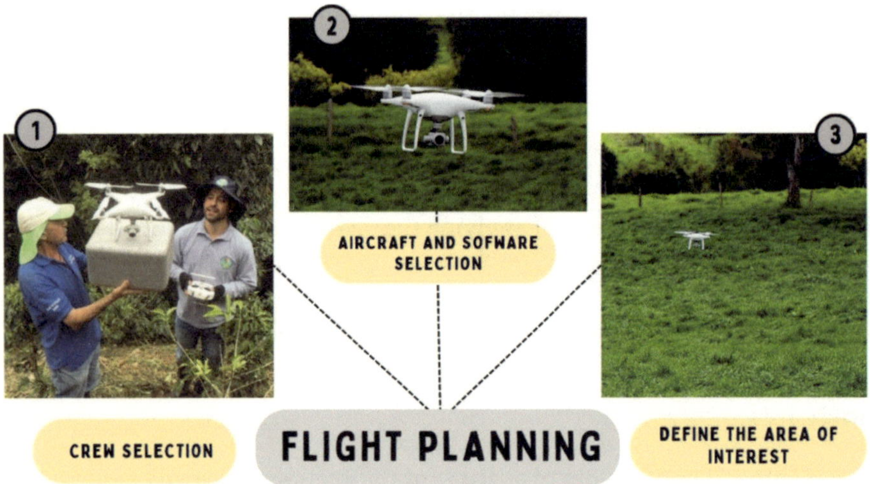


**Figure 2.**
*Steps for UAV flight planning.*

while the visual observer plays a critical role in identifying potential airspace risks, maintaining detailed flight records, and assisting the pilot with equipment handling and calibration. The data manager, on the other hand, is responsible for managing the storage and transfer of information between flights, as well as assessing the quality of the images captured. For successful and safe flights, it is crucial to have a cooperative team that determines the objectives of the operation in advance.

### *2.2.2 Aircraft and software selection*

Aircraft and flight planning software are critical to any aerial project. To achieve stable and effective flights, elements such as battery life and the availability of a stabilizer must be carefully considered. In addition, software must be selected that ensures the recording of areas and flight parameters, allowing autopilot flights under human supervision. This overall strategy ensures the efficiency of data collection and the successful completion of the previously drawn plan. Among the most common software programs for UAV control are DroneDeploy©, DJI Terra©, Litchi©, AirMap©, and Pix4Dcapture© (as part of the Pix4D suite). These programs vary in approach and features, offering options suited to different needs and levels of experience.

Some of these programs are simple and do not require advanced knowledge. They offer the ability to plan automated flights and generate basic 3D maps and models. Others are specific to working with DJI drones, ensuring seamless integration with DJI equipment. Some programs offer more manual control over flights, allowing more detailed customization of the drone's routes and actions. To maintain the safety and legality of flights, the planning and controlling of them need to factor in airspace regulations and constraints.

Programming the flying software requires a suitable flight plan. The AUV must fly directly over the region of interest, thus setting the flying perimeter. Additionally, monitoring flight statistics estimated by the flight software is required. An average flight generates 150–200 photos. The drift angle can be modified to reduce flight time and adjust the flight paths start/end point. It is recommended to start the flight at a greater distance and end the flight closer to the starting point. Since image capture may result in some loss of resolution, it is advisable to maintain a target resolution. In this sense, a resolution of 2–3 cm is recommended for sufficient quality of the processed images, thus working with the machine learning algorithms.

In this case, PIX4Dcapture software (©Pix4D SA-2023) [17] was used for flight preparation. PIX4Dcapture is a tool dedicated to both flight mission organization and photogrammetry. It allows for the meticulous planning of flight paths, allowing the capture of images from different angles. The software additionally offers real-time tracking, automatic pre-launch inspections, and accurate image capturing (**Figure 3**).

Once the pre-flight procedures are completed, it is necessary to check the antenna alignment before starting the propeller motors. Take-off should be executed under manual control to prevent the collision of the drone during flight. The pilot and the visual observer must keep regular visual contact. When landing, if an automatic landing is decided, it is essential to monitor the behavior of the AUV during descent. Promptly rectify any aberrant behavior such as the drone's misguided landing attempts. If on the other hand, a manual landing is required, the drone should be carefully and slowly guided to the desired location. Once on the ground, it is important to download the data collected during flight and shut down the drone. Integration with Pix4D provides a complete workflow from planning to the advanced processing and analysis of captured data, thus, making it easy to achieve accurate and detailed results.

### *2.2.3 Define the area of interest*

The process of selecting the area of interest (AOI) is critically important due to the objectives of the assessment, with a focus on ensuring accessibility to the area of the study. An effective strategy for delineating this terrain is to create a polygon that outlines the boundaries of the AOI using a GPS system. Once established, the polygon can be uploaded to PIX4Dcapture software in KMZ format. This allows flights to be planned within the specific area of interest, resulting in a more accurate and efficient use of flight resources (**Figure 4**).

In this study, areas of interest were selected where the drone images would be taken. As shown in **Figure 5**, a 56.5 cm diameter circular frame with a prominent color was used to distinguish the area of study in the drone imagery. A white and

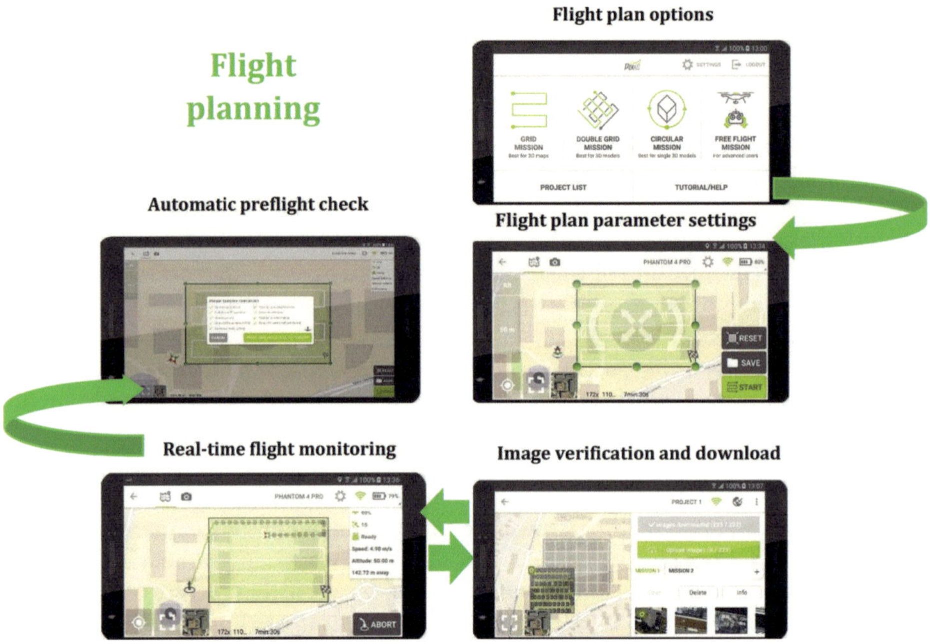


**Figure 3.**
*Flight programming in PIX4Dcapture software.*

**Figure 4.**
*Delimitation of the area of interest in PIX4Dcapture.*

yellow 130 × 130 cm square frame was also used. Alternatively, high-contrast colored spray markings or weather-resistant rubber and vinyl markers can be used ©DroneDeploy-2017 [18]. Another option is to use a lawn frame, which can be clearly seen from flight altitude.

Prior to operating the AUV, the several recommendations considered included weather conditions, reflective panel, and AUV elements:

**Figure 5.**
*Circular and square frames to differentiate the flight and interest area.*

### *2.2.4 Weather conditions*

Ensuring the secure functionality of drones necessitates the presence of consistent and stable weather conditions. Elevated precipitation levels, strong winds, and low visibility could compromise the safety of drone operations. Moreover, dense cloud cover can also block sunlight, resulting in a loss of image quality. For optimum resolution, it is recommended to plan drone operations between 11:00 am and 1:00 pm, when the sun reaches its zenith, ensuring favorable lighting conditions.

### *2.2.5 Reflective panel*

The use of a reflectance panel at the beginning and end of each flight is required to ensure that remote sensing applications are calibrated and ensure accurate alignment with the camera's response to light.

### *2.2.6 AUV elements*

It is essential to check the AUV battery charge prior to flight. It should be noted that the charging process can take up to one hour. For the Phantom 4 Pro UAV, flight time should not exceed 25 min, due to its battery capacity and design limitations. Having two high-capacity SD cards compatible with the UAV will facilitate the organization of the images and ensure that the expected quantity and quality of the captured images are obtained. In terms of flight altitude, it is suggested to keep the AUV at an altitude of approximately 120 m and temperature conditions ranging from 0 to 40°C. This altitude is one of the standard conditions for most flights and provides an adequate balance between flight time and resolution.

## 2.3 Using image processing to create orthomosaics

Photogrammetry uses sophisticated methods for creating and capturing accurate three-dimensional data from images. Once the files have been downloaded from the UAV, the steps that follow in Agisoft PhotoScan Software, are described in **Figure 6**.

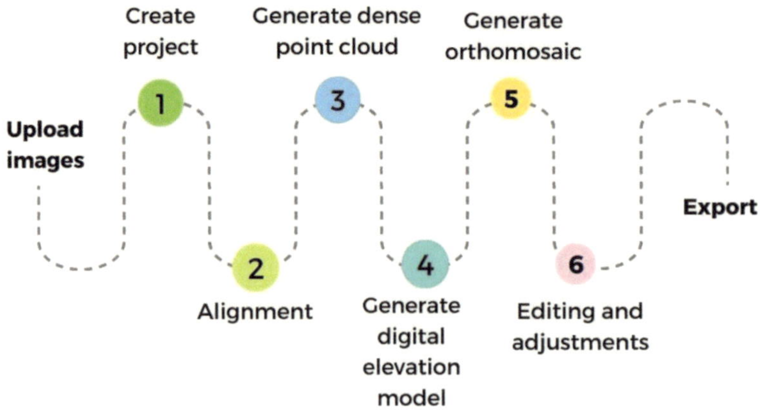


**Figure 6.**
*Steps for image processing in Agisoft PhotoScan software©.*

The images captured by the drone are imported into the processing software. Once imported, the images undergo a standard processing procedure to eliminate distortions and rectify variations in lighting. At this point, the geographical coordinates of each image are also established to ensure precise spatial positioning. The next step is to orientate the images, aligning them in a coherent and precise manner to achieve proper alignment between them all. Once this phase is completed, the generation of a three-dimensional point cloud from the processed images is initiated. This point cloud contains detailed information about the terrain's topography and the structure of the objects in the scanned area, which is critical for obtaining accurate and complete representation of the environment.

The information in the point cloud is used to create the orthomosaic. This composite image combines all the individual images into an accurate panoramic view of the area, representing the terrain and objects in detail and with accuracy. The resulting orthomosaic offers a complete and precise geospatially visual depiction of the area of interest and can be utilized in a range of applications, including mapping, urban planning, and environmental monitoring. Together, these steps form a comprehensive process that transforms the original imagery into a visually coherent and geospatially accurate representation of the terrain and its surroundings, using PhotoScan software as a key tool throughout the process. For this study, 18 images were taken at each of the selected properties to obtain different RGB values and allow for correlation between them and the bromatological results of the pastures. The grass species found in these properties were mainly varieties of Urochloas, native grass, kikuyu grass and star grass. Image processing and sampling point identification is shown in **Figure** 7.

## 2.4 Collection and pre-processing of RGB images

The RGB images generated in the flights of the rural properties correspond to the total area of the property (if it is small). The images were taken at a height of 60 m with an overlap of 90% and stored in JPG format. The RGB band values from the images were extracted using Google Colab *via* the Python programming language.

The *cv2, numpy, matplotlib,* and *pandas* libraries were used to import the images, select the region of interest, and extract the RGB values. The region of interest (ROI) was selected by defining coordinates in the center of each image, considering it to be a

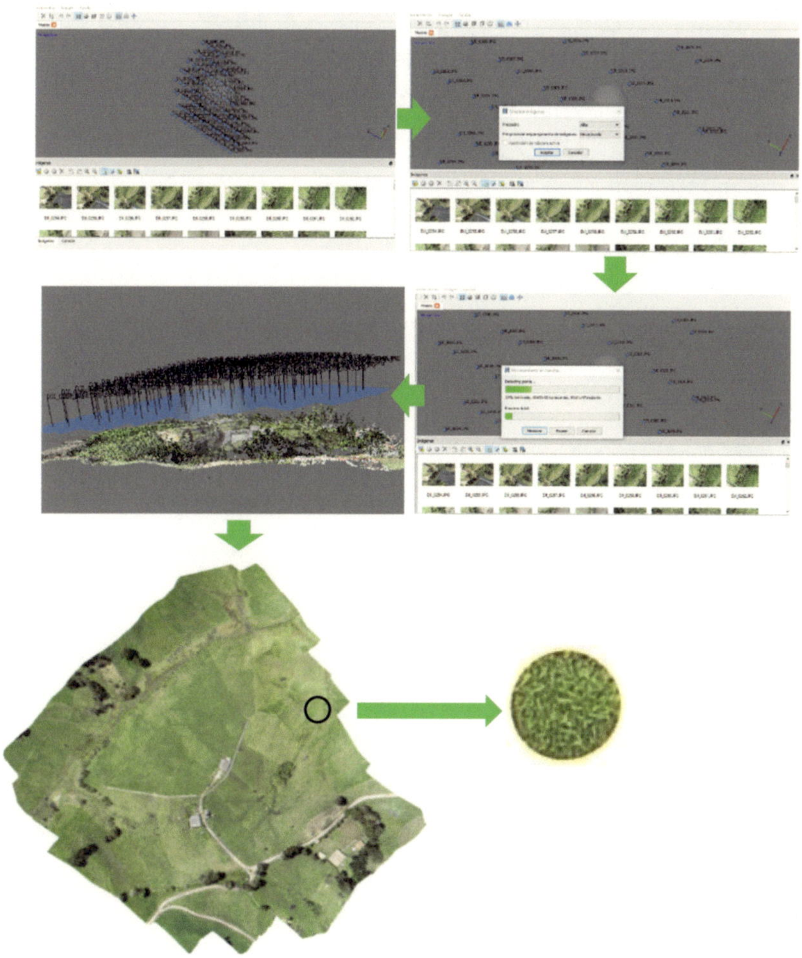

**Figure 7.**
*Drone footage capturing pastures.*

homogeneous area; from this region of interest, the average values of the RGB bands were extracted. The steps to perform the pre-processing of the images in Google Colab are shown in **Figure 8**. The purpose of pre-processing the images is to obtain three numerical values of the RGB bands that will be applied to the statistical models.

The averages of each band for each of the attributes were used to obtain the RGB values. The values of the RGB bands can be in the range or scale from 0 to 255. The results are presented in **Table 2**.

## 2.5 Bromatological analysis of grass

The pastures were sampled in the field following the methodology guidelines for bromatology sampling established by the Department of Research Laboratories and Services of Agrosavia. The crude protein (PC), ethereal extract (EE), total moisture (H), neutral detergent fiber (NDF), and acid detergent fiber (ADF) contents were determined by wet chemical analysis. The values reported by the bromatological analysis of each farm were included in the database with the RGB values of each image. This

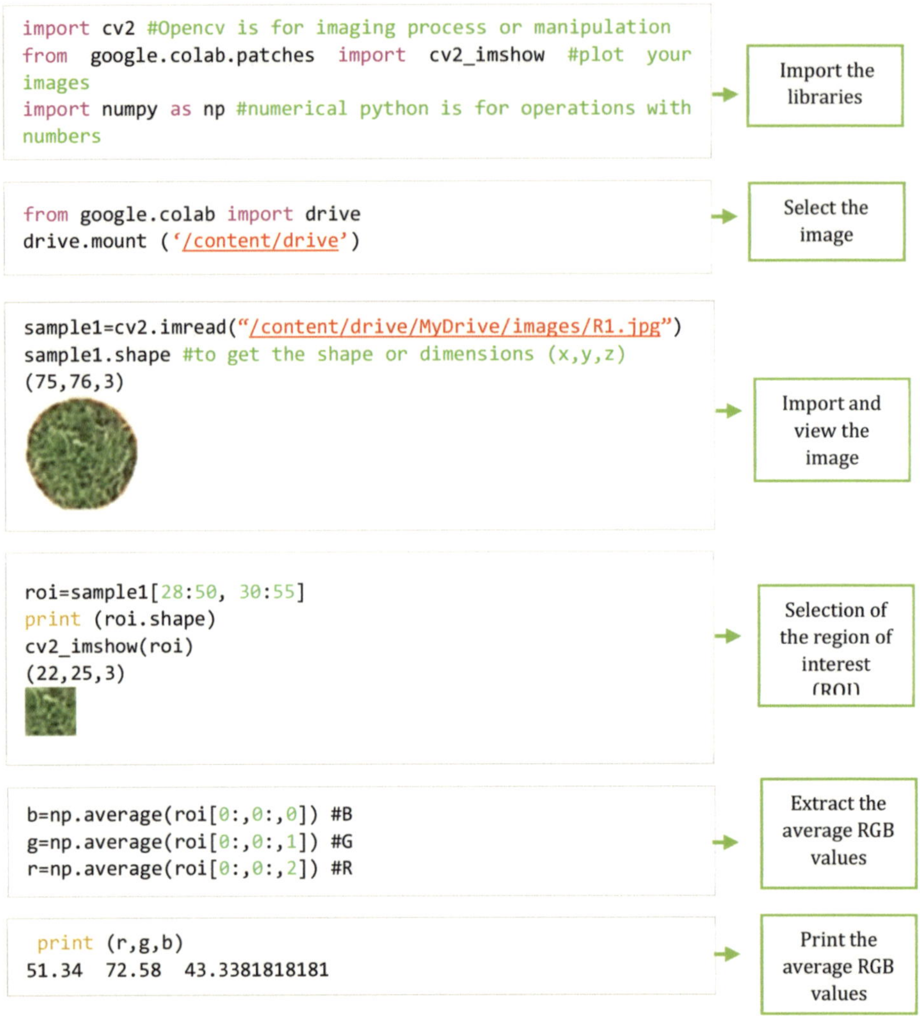


**Figure 8.**
*A flowchart indicating the Programming of RGB image pre-processing done in Google Colab.*

database was used in the R-Project statistical software to apply statistical regression models and predict the nutritional characteristics of the pasture using the RGB bands.

## 2.6 Statistical processing

Statistical processing of the data was performed in the software R-Project [4.2.2]. The "*gramEvol*" library [19] was used to generate different indices with the RGB bands using grammatical context-free expressions with minimum cost functions. The logarithm, exponent, square root functions, and the "+", "-", "*", "/" operations were included; in addition, the logarithmic transformation was applied to the pasture nutrient variables. However, the models showed a better fit when the transformation was not applied to the data.

In addition to the indices obtained with *gramEvol*, another 50 indices and their combinations were developed to find the best equations. The equations had to meet

| Municipality | R band | G band | B band |
|---|---|---|---|
| Sopetrán | 130.28 | 156.33 | 96.97 |
| Sopetrán | 149.28 | 164.61 | 118.22 |
| Maceo | 132.16 | 153.47 | 80.79 |
| Abejorral | 119.66 | 145.31 | 79.32 |
| Abejorral | 79.98 | 140.99 | 65.13 |
| Yarumal | 109.42 | 155.06 | 76.74 |
| Yarumal | 74.10 | 121.69 | 51.32 |
| Amagá | 129.99 | 150.60 | 89.56 |
| Amagá | 115.16 | 154.26 | 95.95 |
| Santa Rosa de Osos | 99.77 | 151.74 | 55.10 |
| Santa Rosa de Osos | 101.94 | 141.62 | 73.83 |
| Chigorodó | 132.07 | 149.83 | 56.84 |
| Chigorodó | 147.55 | 174.31 | 80.46 |
| Caucasia | 160.00 | 198.43 | 118.79 |
| Caucasia | 131.39 | 169.25 | 100.15 |
| Támesis | 145.30 | 180.33 | 82.93 |

**Table 2.**
*Average RGB bands extracted from the images of each farm in the study.*

| Variable | Best model | $R^{2*}$ |
|---|---|---|
| Acid detergent fiber (ADF) | $y=14.79e^{\frac{R}{G}}$ | 0.99 |
| Moisture (MOISTURE) | $y=9.15\frac{R}{G}-1.94e^{\frac{R}{B}}$ | 0.97 |
| Crude protein (CP) | $y=-63.36e^{\frac{R}{G}}+173.88\sqrt{\frac{R}{G}}$ | 0.95 |

*$^*R^2$ = coefficient of determination.*

**Table 3.**
*Equations and fit obtained by means of linear models for the variables ADF, CP, and MOISTURE.*

the criteria of highest $R^2$ and best Bayesian Information Criterion (BIC) for the nutritional variables. Of all the nutritional variables analyzed, good indices were obtained only for ADF, CP, and MOISTURE. The results are shown in **Table 3**.

Authors such as [20] employed various indices proposed by other researchers in their studies, enhancing the precision of results in images captured with RGB cameras. Furthermore, authors such as [10] used indices based on NDVI values, creating different combinations between this index and the RGB bands; with which they were able to obtain adjustments of 0.71 in maize crops.

## 3. Conclusions

The application of a well-studied field methodology for RGB image acquisition in pastures can mitigate errors before, during and after drone flights, providing greater accuracy and efficiency throughout the process. In addition, the equations constructed using the RGB band indices were presented an excellent fit and can be used to predict pasture nutritional variables in future pasture research. The example presented in this chapter will facilitate image acquisition and information processing, as it can be extrapolated to similar studies. However, a more substantial quantity of RGB data is recommended for data processing and statistical modeling and the insufficient sampling size stands out as a noticeable constraint. Increasing the sample size in data collection enhances the validity of research findings.

From a positive perspective, the utilization of RGB images proved beneficial by enhancing decision-making efficiency over extensive crop areas. In addition, the methodology, curated and carefully selected from previous studies, allows for easy replication of the technique in various locations. While these techniques demonstrate effectiveness in analyzing pasture crops, their applicability extends to diverse crops such as corn, rice, and avocado, among others. This tool can be used by government entities providing technical assistance to producers for general livestock feeding plans. Additionally, in certain instances, it can function as an adjunctive support to pasture fertilization plans.

In addition to RGB images taken by UAV, there are other more complex techniques such as multispectral images (MSI) and hyperspectral images (HSI), which can provide more precise information about the nutritional conditions of crops due to the greater number of bands they encompass compared to RGB. However, it is imperative to note that these methods necessitate further research and field testing for comprehensive validation.

## Acknowledgements

This work received financial support for the fieldwork from the project "Development and Establishment of the Centre for Agro-biotechnological Development of Innovation and Territorial Integration, El Carmen de Viboral, Antioquia, Occident (CEDAIT)", Expert System Component, with resources from the General Royalty System and the Government of Antioquia. Financial support was also received for the publication of this chapter from the project "Design and validation of predictive models to determine Cation Exchange Capacity (CEC), Organic Matter (OM) and Nitrogen (N) in soils from hyperspectral images" through the agreement 2022-7204, financed by the University of Antioquia Foundation. The authors express their gratitude to Tatiana Rodríguez Monroy, a master's student in Animal Sciences, for her contributions.

## Conflict of interest

The authors declare no conflict of interest.

## Author details

Valentina Parilli-Ocampo[1], Manuela Ortega Monsalve[2], Mario Cerón-Muñoz[3], Luis Galeano-Vasco[3] and Marisol Medina-Sierra[3*]

1 Earth Sciences, University of Glasgow, Scotland

2 Faculty of Agrarian Sciences, Animal Science, GAMMA Research Group, Antioquia University, Medellin, Antioquia, Colombia

3 Faculty of Agrarian Sciences, GAMMA Research Group, Antioquia University, Medellin, Antioquia, Colombia

*Address all correspondence to: marisol.medina@udea.edu.co

## References


[1] Eissa AHA, Khalik AA, Abdel AA. Understanding color image processing by machine vision for biological materials. In: Eissa AHA, editor. Structure and Function of Food Engineering. Chapter 10. Intechopen; 2012. pp. 227-274, 416 p. DOI: 10.5772/50796

[2] Pharr M, Wenzel J, Humphreys G. Physically Based Rendering: From Theory to Implementation. 4th ed. MIT Press; 2023. 1312 p. Available from: https://pbr-book.org/4ed/Radiometry,_Spectra,_and_Color/Color

[3] Fu L, Gao F, Wu J, Li R, Karkee M, Zhang Q. Application of consumer RGB-D cameras for fruit detection and localization in field: A critical review. Computers and Electronics in Agriculture. 2020;**177**:105687. Available from: https://www.dronedeploy.com/blog/what-are-ground-control-points-gcps/

[4] Torres-Sánchez JM, Peña AI, de Castro A, López-Granados F. Multi-temporal mapping of the vegetation fraction in early-season wheat fields using images from UAV. Computers and Electronics in Agriculture. 2014;**103**:104-113. DOI: 10.1016/j.compag.2014.02.009

[5] Puri V, Nayyar A, Raja L. Agriculture drones: A modern breakthrough in precision agriculture. Journal of Statistics and Management Systems. 2017;**20**(4):507-518. DOI: 10.1080/09720510.2017.1395171

[6] Li W, Niu Z, Wang C, Huang W, Chen H, Gao S, et al. Combined use of airborne lidar and satellite GF-1 data to estimate leaf area index, height, and aboveground biomass of maize during peak growing season. IEEE Journal of Selected Topics in Applied Earth Observations and Remote Sensing. 2015;**8**(9):4489-4501. DOI: 10.1109/JSTARS.460944310.1109/JSTARS.2015.2496358

[7] Luna I, Lobo A. Mapping crop planting quality in sugarcane from UAV imagery: A pilot study in Nicaragua. Remote Sensing. 2016;**8**(6):1-18. DOI: 10.3390/rs8060500

[8] Deshmukh S, Jadhav V. Bromatological and mineral assessment of Clitoria ternatea linn. Energy (KJ). 2014;**459**(9):489-424

[9] Manzani L, Aparecido L, Ventura G, Ferreira L, Monteiro de Figueiredo PA. Bromatological and morphological characteristics of forage plants. Investigación Agraria. 2021;**23**(1):22-27. DOI: 10.18004/investig.agrar.2021.junio.2301602

[10] Vergara-Díaz O, Zaman-Allah MA, Masuka B, Hornero A, Zarco-Tejada P, Prasanna BM, et al. A novel remote sensing approach for prediction of maize yield under different conditions of nitrogen fertilization. Frontiers in Plant Science. 2016;**7**(666):1-13. DOI: 10.3389/fpls.2016.00666

[11] Karila K, Alves R, Ek J, Kaivosoja J, Koivumäki N, Korhonen P, et al. Estimating grass sward quality and quantity parameters using drone remote sensing with deep neural networks. Remote Sensing. 2022;**14**(11):2692. DOI: 10.3390/rs14112692

[12] Barbedo JG. Digital image processing techniques for detecting, quantifying, and classifying plant diseases. SpringerPlus. 2013;**2**(1):1-12

[13] Tagle Casapía X, Di Liberto S, Falen L, Flores G, Dávila A, Mendoza C,

et al. Protocolo para sobrevuelos con RPAs PHANTOM 4 PRO y PHANTOM 4 RTK. 1st ed. Perú; 2021. Available from: https://repositorio.iiap.gob.pe/bitstream/20.500.12921/610/5/tagle_protocolo_2021.pdf

[14] Posada-Asprilla W, Cerón-Muñoz MF. Influencia del ángulo de iluminación solar y la altura de la toma de la imagen multiespectral sobre la estimación de biomasa de pasto kikuyo. Revista UDCA Actualidad & Divulgación Científica. 2019;**22**(2):1-6. DOI: 10.31910/rudca.v22.n2.2019.1338

[15] Toskov B, Toskova A, Stoyanov S, Doychev E. Architecture of intelligent guard system in the virtual physical space. In: Sgurev V, Jotsov V, Kruse R, Hadjiski M, editors. IEEE 10th International Conference on Intelligent Systems (IS); August 28-30; Sofia, Bulgaria. 2020. pp. 265-269, 629 p. DOI: 10.1109/IS48319.2020.9200177

[16] Pajares G. Overview and current status of remote sensing applications based on unmanned aerial vehicles (UAVs). Photogrammetric Engineering & Remote Sensing. 2015;**81**(4):281-330

[17] Pix4Dcapute. Un paquete de software para mapeo. 2023. Available from: pix4d.com/es/

[18] DroneDeploy. Online: How Do I Use Ground Control Points? 2017. Consulted on August 15, 2023. Available from: https://www.dronedeploy.com/blog/what-are-ground-control-points-gcps

[19] Noorian F, de Silva AM, Leong PHW. gramEvol: Grammatical evolution in R. Journal of Statistical Software. 2016;**71**(1):1-26. DOI: 10.18637/jss.v071.i01

[20] Barbosa BDS, Ferraz GAS, Gonçalves LM, Marin DB, Maciel DT, Ferraz PFP, et al. RGB vegetation indices applied to grass monitoring: A qualitative analysis. Agronomy Research. 2019;**17**(2):349-357. DOI: 10.15159/ar.19.119

Chapter 5

# Big Data Analytics for Yield Prediction in Precision Agriculture

*Vasudevan N. and Karthick T.*

## Abstract

India's agriculture industry is crucial to its economic growth and one of its most fundamental endeavors. Regarding a country's economic prosperity, agriculture is among the most significant factors contributing to the happiness and well-being of its citizens. To improve agricultural output, "smart agriculture," or the use of technology, strives for more accurate disease control, irrigation, and yield prediction. Precision agriculture is applying big data analytics and the Internet of Things to the farming industry. Agricultural production will increase dramatically as a result of this. The Internet of Things (IoT) and massive amounts of data are used in precision agriculture to improve crop quality and yields. In this research, we use the grape plants and their associated factors (temperature, humidity, rainfall, pH, sun irradiance, etc.) from the Smart Agriculture dataset to develop an N-stage CNN. In this work, we use machine learning approaches for irrigation scheduling and the DoubleGAN methodology for disease diagnosis in plants. This effort aims to create an N-stage CNN model that will significantly boost agricultural output by enhancing grape plant yield. The yield prediction is quite accurate since we considered practically all necessary characteristics and photos for plant development, including irrigation schedule and disease detection.

**Keywords:** precision agriculture, CNN, GPU, smart farming, IoT, big data analytics

## 1. Introduction

Food production must grow by 70 percent in the year 2050 in order, according to Beecham's paper entitled "Towards intelligent farming: agriculture that embraces IoT vision," to meet our anticipated world population of 9.6 billion persons. Commercial markets are coming up nowadays; there is also a need for industrialization, and construction of residential buildings is taking place in agricultural fields, and these can be directly blamed on the increase in population, which in turn requires a rise in food production.

In simple words, IoT can be explained as an infrastructure of interconnected objects, people, or systems to exchange and connect information with other systems and devices over the online network. It is a system of interrelated digital machines, computing and mechanical devices, objects, and people. Implementing this interaction between a human and a human, as well as between a computer and a human, comes into action. In today's world, seamless communication is possible through

 IntechOpen

processes, things, and people. This is because of the evolution of the enormously growing technology Internet of Things. These pre-built Software-as-a-Service (SaaS) applications have made lives more accessible in the modern age. Various kinds of applications, data, and sensors are being connected here. Some notable applications are self-driving cars, smart microwaves, fitness devices, connected footballs, voice controllers, smart locks, and many others.

When there are massive datasets, analytical techniques are implemented, called Big Data Analytics. The advancement of Big Data incorporates various primer strides for its establishment. "Big Data" is a relative term depending on who discusses it and for what purpose. This includes semi-structured, unstructured, and structured data from various sources. This might also include massive amounts of datasets ranging from terabytes to zettabytes. Structured data sets are collections of data that can be utilized in their original form and from which results may be obtained. Unstructured data sets are the opposite of structured ones, without proper formatting and alignment. Relational data, such as employee salary records, comes under structured data, and human texts, Google search result outputs, etc., come under unstructured data. When there is a combination of both structured and unstructured data, it is called a semi-structured dataset. These data sets lack defining elements for sorting and processing, yet, they might have a proper structure. RFID and XML data are generally semi-structured data.

The quality of crops is essential for increased production, which also requires yield prediction. Observing the plant and disease detection is also necessary for determining the plant growth and, thereby, the total yield. Almost all the properties needed for the development of plants will be considered, and diseases also will be detected as a supporting factors. Previous works have concentrated only on two or three growth parameters, and some others have only devised methods for disease detection. In this paper, we have considered growth parameters as well as the detection of disease. Nowadays, farming is not passionately taken forward by farmers due to various issues such as insect attack diseases, sudden changes in temperature, climate, irregular rainfall, and many others. Technology advancements are required to make farming more friendly, profitable, and intelligent. There are great opportunities for discovery and innovation in precision agriculture. We have surveyed and researched extensively and used the most efficient methods to provide maximum yield.

In this work, we have immensely used the Internet of Things and Big Data Analytics methods to increase crop production. The application of these technologies in an agricultural field is called Precision Agriculture System. Big data and IoT are the two significant on which our model relies. These will then be combined by Precision Agricultural System using n-stage CNN, and we will devise a smart farming method, which will provide maximum yield.

CNN is a deep learning technique that is used to solve complicated issues. CNN is the most common approach utilized. It surpasses standard methods of machine learning [1, 2]. An automated CNN design approach can be formed by employing evolutionary algorithms to construct photo-classification jobs successfully. The most significant advantage of this method remains its "automated" feature that users do not need CNN domain expertise yet have an excellent CNN architecture for the pictures [3]. To make the training of networks that are far deeper than those used earlier, the DoubleGAN technique using VGG16 and the Residual learning framework was utilized [4, 5].

In contemporary farming management, water supply distribution plays an essential role. Irrigation controls based on evapotranspiration can ideally supply irrigation in line with the plant's water needs. A predictive irrigation schedule was employed in nurseries with various crops and high-frequency water requirements under limited resources. The time-series analysis is vital to the water balance equation using historical data to calculate evapotranspiration. A hierarchical research algorithm was proposed to predict irrigation schedules to minimize crop stress periods and optimize resources. This algorithm includes the dispatch of priority regulations and the accounting for crop characteristics, available water, and hydraulic system constraints [6]. Irrigation plans can consist of real-time input on soil humidity and climate sensors. However, models of the dynamism of soil humidity are necessary for robust closed-loop decisions to anticipate crop water requirements while responding to outside perturbations and troubles [7].

One of the most critical tasks in agriculture is the identification and categorization of plant diseases by using appropriate imaging and machine learning technology. Through the proper detection of illnesses, we can manage them. This increases crop production automatically.

The methods used to diagnose plant illnesses from the year 2015 to the year 2017 include Support vector machines [8], Delta color difference algorithms, Histograms and text characteristics [9], Sparse Representation [10], and Convolution neural networks [11].

The neural networks significantly affected plant disease recognition between 2018 and 2019. During those eras GoogleNet [12], radial-based functions [13], VGG16 (Visual Geometrical Group) [14], MCNN (Multilayer CNN), ResNet, etc. were the most often utilized neural-network techniques [15–17]. The current detection of plant disease includes generative adversarial Networks, Extended ROI (Region of Interest), LSTM, and Inception V3 [18–20]. LeafGAN, DATGAN, and DoubleGAN are the most utilized neural network techniques using GAN [21–23].

Machine learning is a way to gather and publish enormous volumes of data. The agronomic concepts of crop modeling were integrated with machine learning to develop a machine-learning basis for a broad crop output projection. The primary line is an accuracy, modularity, and reusability procedure. For correction, we can concentrate on creating and using machine learning without information leakage, explainable predictors, or characteristics in connection to crop growth and development [24]. ML with system vision is monitored to categorize several crop images to observe the crop quality and the output valuation. This technique may be combined into improved livestock production by predicting reproductive patterns, diagnosing eating problems, animal behavior using ML models, etc. [25, 26].

The primary need in agriculture is to develop an effective model using machine learning algorithms and neural networks for the maximum yield. And the model should fulfill the following needs helping farmers to produce a total yield, early detection of diseases before it becomes fatal, and finding appropriate irrigation duration.

The stated objective of this article is as follows:

- We are developing a methodology that uses Convolutional Neural Networks to forecast the optimal production yield. The variables considered for prediction include temperature, humidity, rainfall, and sun irradiation.

- Furthermore, we are now developing a technique for illness diagnosis using the DoubleGAN approach. This approach enables more efficient and exact yield prediction than other methods relying only on certain metrics or illness detection.

## 2. Dataset and study area description

Machine learning models require an enormous volume of data to be processed efficiently. Adaptable information simplifies the effort to identify regularities by eliminating unnecessary features for learning. Developing a deep-reinforcement learning model for the agricultural framework is exceedingly laborious as it is a highly linear dynamic behavior. This section details the data set utilized for irrigation and agrarian yield prediction. The study examines the predictive yield of grape plant land in southern India for the district of Salem. The dataset contains particular climate, soil, and soil characteristics and the number of fertilizers consumed by the grape plants in the research field, according to the typical meteorological and soil factors.

The data collected for the irrigation phase included temperature, humidity, precipitation, and pH of grape crop type. The grape plant leaves picture data sets was obtained for plant disease identification and agricultural yield forecast. Data are collected for 15 years. Knowledge has been used about regular climate variables, such as temperature, precipitation, crop evapotranspiration reference, potential evapotranspiration, humidity, and character distinct characteristics, such as ground frost frequency and diurnal and wind speeds. The Indian Meteorological Department supplies climate data and grape plant images through their metadata tool webpage. The soil characteristics include the highest soil density and PH and the number of soil macronutrients (nitrogen, phosphorus, and potassium). The following section illustrates how the technique proposed for these datasets is utilized.

## 3. Methods and materials

The proposed work is divided into the Irrigation phase, plant disease detection phase, and crop yield phase. The irrigation phase output is compared using the machine learning methods naïve Bayes algorithm, Random forest tree algorithm, and decision tree algorithm. The plant disease is identified using the DoubleGAN method, and the final crop yield is predicted using the aggregation of the irrigation phase, plant disease detection, and crop yield prediction (**Figure 1**).

The above figure shows that the sensor data which n-stage CNN segmented. It had two phases, the first phase was the irrigation phase, in which we had the input taken

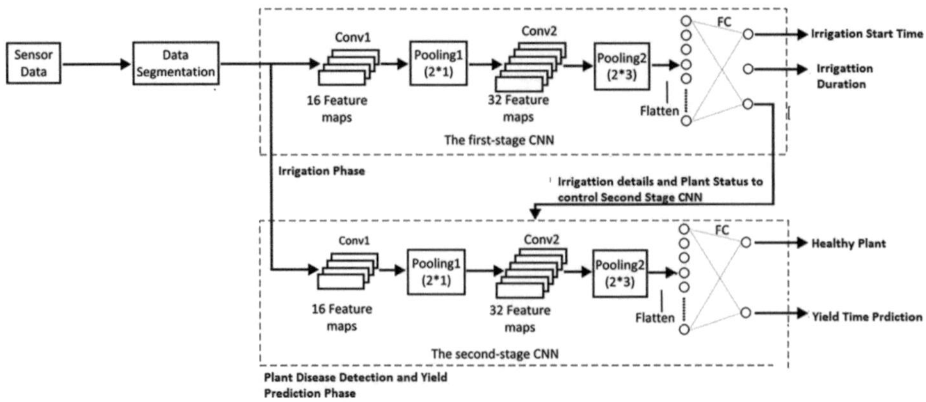


**Figure 1.**
*System architecture.*

to a convoluted layer, and then it was shortened in size by pooling. And this process continued until the system reached got fully connected layer. From this, two outputs are formed and sent to the second phase, the disease detection phase. Again, the same process happened in the second phase, and we got two results: the healthy plant and yield prediction, which is our final output. The algorithm of the proposed work is given below.

### 3.1 Irrigation

The irrigation phase is the first phase in our convolutional neural network. In this phase, datasets had collected such as temperature, humidity, rainfall, pH, and nature of crops. The n-stage CNN segments these datasets. At the end of this phase, two outputs formed: irrigation start time and duration. In this, the two outputs, the irrigation start time and duration, are static, and the irrigation details and plant images are sent to the second phase, plant disease detection. So this is the overall methodology of our first phase. To implement the irrigation phase, three algorithms were used with grape plants. They are the decision tree, random forest, and naive Bayes algorithms, and these algorithm results are compared.

### 3.2 Disease detection

First of all, the image datasets are collected. Now let's see the steps involved in the CNN implementation for the disease detection phase. The output, the irrigation details from the first phase, and plant leaf images were taken as the input for this phase. First, the input was information to the convolutional layer and then moved to the pooling layer, which reduced its size to half. Then the same process continued until it reached the fully connected layer. From this phase, two outputs formed: the healthy plant and yield prediction, which is our final output. The DoubleGAN method is used for image processing as an augmentation tool [27].

Plant leaves can be used to identify infection in plants. However, there are generally uneven depictions of uneasy leaves taken from different plants. Diseases with such an imbalanced dataset are challenging to detect. We employed DoubleGAN (a double generative adversarial Network) to equilibrate such data sets. We suggested utilizing DoubleGAN to create high-resolution pictures with fewer samples of sick leaves (**Figure 2**).

The two steps of Double GAN are split. As ingredients in stage 1, we used healthy leaves and unhealthy leaves. Firstly, the pre-trained model was supplied with the WGAN using healthy photos (Wasserstein generative opposing network). The model then used sick leaves to create sick leaves of 64*64 pixels. In the second stage, 256 × 256-pixel frames were obtained to extend unbalanced data settings utilizing an

**Figure 2.**
*(a): DoubleGAN first stage generation grape leaf image; (b): DoubleGAN second stage generation grape leaf image.*

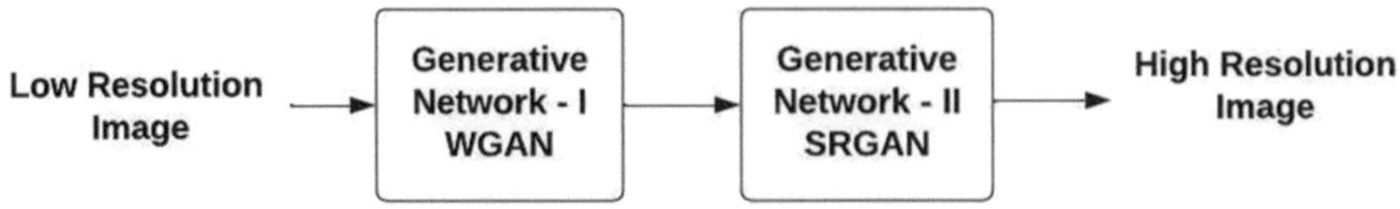


**Figure 3.**
*DoubleGAN architecture.*

adversarial network in super-resolution (SRGAN). Finally, the DCGAN images were compared (Deep convolution generative adversarial network) (**Figure 3**) [24].

### 3.3 Crop yield prediction

The forecast of crop production and how yield may be enhanced is essential knowledge for every farmer. The pH value, soil type, and quality are critical in providing crop yields: temperature, precipitation, moisture, sunlight hours, fertilizers, and harvest schedules [28]. Manual farming may be viewed scientifically as a feedback control system where remedial measures can be taken when a reverse effect is seen in a crop. Crop production depends heavily on the effectiveness of the best use of these resources. In the first step, if any abnormality is not recognized, the agricultural production might be damaged without precedent [29].

For crop monitoring and estimating/prediction of yield using remotely sensed data, machine learning (ML) approaches are used. The difficulties with harvest forecast methods using time series of distant information might lead to discrepancies between the fields of the standard index of vegetation changes generated from satellite (NDVI). This study employed the advanced approaches used for ML, such as boosting regression tree (BRT), random forest regression (RFR), and support for the reversal of vectors and Gaussian GPR. The normalized vegetation differential index (NDVI) measurements were averaged and integrated each year for each field to generate a double-dimensional data set. The ML methods have been utilized 100 times, and their assessment measures have been used to assess performance and study stability. Finally, every ML method has averaged its findings to produce returns. Comparisons between these approaches reveal that the average R-value of BRT throughout all years is over 0.87 [30].

The NDVI is a simple graphical indicator for the analysis of distant sensing measures, frequently from a space platform, which assesses whether the target is seen containing living green vegetation. This is a standardized difference vegetation index. The vital forces of the two algorithms are combined via boosted regression trees. It's a regression tree and growing. The regression tree relates a reaction via recurrent binary splits to its predictors. The boost is an adaptive way to combine multiple accessible models to enhance prediction performance.

### 3.4 Algorithm: precise agriculture system

1. *Input : H,R,T,C,pH,SI, Images*

   Where $H$ denotes Humidity,
   $R$ denotes Rainfall,
   $T$ denotes Temperature,
   $C$ denotes Nature of Crop,

*pH* denotes Power of Hydrogen,
*SI* denotes Solar Irradiance.

2. *Output: IS, ID, PS, YT*

   Where *IS* denotes Irrigation
   Start Time,
   *ID* denotes Irrigation
   Duration,
   *PS* denotes Plant Healthy
   Status,
   *YT* denotes Yield Time

3. Apply Decision tree, Random Forest and Naïve Method on Inputs *H,R,T,C,pH* and *SI*

4. Choose the method which produce the utput i.e *IS* and *ID* with better accuracy

5. Improve the input images resolution by applying the DoubleGAN method

6. Identify the Plant status i.e plant is defected or healthy

7. Apply Boosted Regression Tree method and predict the yield time

8. Fine-tune the irrigation scheduling, plant disease detection and crop yield prediction output by considering all the applications output.

The precision agriculture system algorithm uses humidity, rainfall, temperature, crop, pH, solar irradiance, and plant images as inputs. The algorithm's output includes the irrigation schedule, plant healthy status, and yield prediction. The precision agriculture system algorithm comprises the irrigation, plant disease, and yield prediction phases. The results and findings of this algorithm are explained in the following section.

## 4. Experimental results and discussion

The proposed method is evaluated on grape leaves with humidity, temperature, rainfall, crop, pH, and soil irradiance to identify the irrigation schedule. Then the image dataset is used to detect plant status and crop yield. The evaluation of the proposed approach is planned through classification methods, including the Decision tree method, random forest method, and naïve Bayes methods. The plant disease detection is performed using DoubleGAN and is evaluated using ResNet and VGG algorithms. Finally, the crop yield prediction is made using Boosted regression tree method.

### 4.1 Irrigation result

The correct irrigation schedule is evaluated using the Decision tree, random forest, and naïve Bayes methods with the parameters Mean Square Error (MSE), Mean Absolute Error (MAE), R-Squared Error, Root Mean Square Error, and accuracy. But it still requires the expert's validation to predict the exact irrigation scheduling.

| Algorithm | MSE | MAE | R-Squared | RMSE | Accuracy (%) |
|---|---|---|---|---|---|
| Decision Tree | 7.63 | 2.16 | −0.87 | 2.77 | 70 |
| Random Forest | 7.51 | 2.14 | −0.82 | 2.74 | 75 |
| Naive Bayes | 10.82 | 2.58 | −1.62 | 3.29 | 65 |

**Table 1.**
*Comparison of ML models on irrigation.*

We can also consider other criteria like plant disease detection and crop yield time to get precise predictions about irrigation scheduling. The following section and **Table 1** show the different machine-learning algorithm performance results for grape leaves irrigation scheduling.

The above graphs represent the Random Forest, Decision Tree, and Naive Bayes metric values. As we can see, among the three algorithms, Random Forest had the highest accuracy, and thus, it proves that this algorithm is the best implementation method for the irrigation phase.

The average or mean of the squares of the expected and predictable target values in a dataset is used to calculate the Mean Square Error.

$$\text{MSE}=1/n^{*}\Sigma\ (e\_i-p\_i)^{\wedge}2 \tag{1}$$

Where e_i is the dataset's i's expected value and p_i is the dataset's i's predicted value.

The mean absolute error is a popular statistic because the error score units, such as RMSE, correspond to the team of the projected target value. The MAE may be calculated using the formula shown below.

$$\text{MAE}=1/n^{*}\Sigma \text{abs}(e\_i-p\_i) \tag{2}$$

Where e_i is the dataset's i's expected value and p_i is the dataset's i's predicted value, and abs() is the absolute function.

The R-squared (R2) statistic shows the variation in a dependent variable that can be explained by an independent variable or variable in a regression model.

The formula below can be used to calculate the RMSE.

$$\text{RMSE}=\text{sqrt}\ \left(1/n^{*}\Sigma(e\_i-p\_i)^{\wedge}2\right) \tag{3}$$

Where e_i is the dataset's i's expected value and p_i is the dataset's i's predicted value.

Accuracy is defined as the percentage of correct test data predictions. The amount of right predictions may be readily estimated by dividing them by the total.

Accuracy can be defined as the proportion of adequately categorized cases.

$$\text{Accuracy}=x/y \tag{4}$$

X is the sum of true positive and true negative, and y is the sum of True Negative, False Negative, False Positive, and True Positive (**Figure 4**).

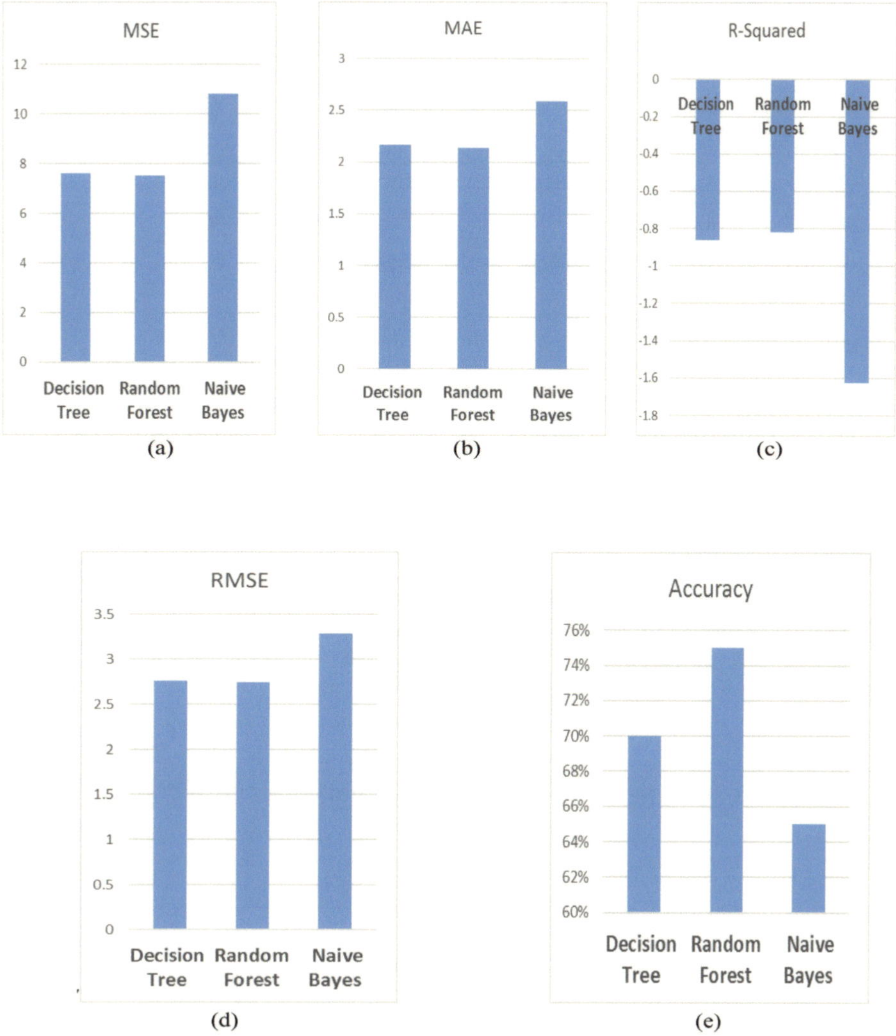


**Figure 4.**
*(a): MSE Metrics; (b): MAE Metrics; (c) R-Squared metrics; (d): RMSE metrics; (e) Accuracy Metrics.*

## 4.2 Plant disease detection results

The DoubleGAN is used to expand the low-resolution images of grape leaves into high-resolution images. And then, the accuracy is evaluated with the VGG16 and resNet50. The results of the various algorithm are shown in **Table 2**.

The results demonstrate that the classifications are better after DoubleGAN expands the data set. VGG16 and ResNet50 are classified very closely in the data sets, so DoubleGAN images are similar to those produced initially. Based on random noise, the DoubleGAN may create multiple images, overcoming the difficulty of insufficient variety generated by the extension of the original data set by spinning and transformation expansion. These plant disease detections also influence irrigation and crop yield time predictions (**Figure 5**).

| Plants | | Tomato | Potato | Corn | Accuracy average |
|---|---|---|---|---|---|
| Original DataSet | VGG16 | 96.45 | 97.65 | 97.48 | 97.19 |
| | ResNet50 | 96.65 | 97.83 | 97.61 | 97.36 |
| Flipping and Translation Expansion | VGG16 | 98.45 | 98.15 | 97.85 | 98.15 |
| | ResNet50 | 98.75 | 98.35 | 97.85 | 98.32 |
| DoubleGAN Expansion | VGG16 | 97.83 | 99.25 | 99.8 | 98.96 |
| | ResNet50 | 98.85 | 99.45 | 99.88 | 99.39 |

**Table 2.**
*Plant disease detection model accuracy.*

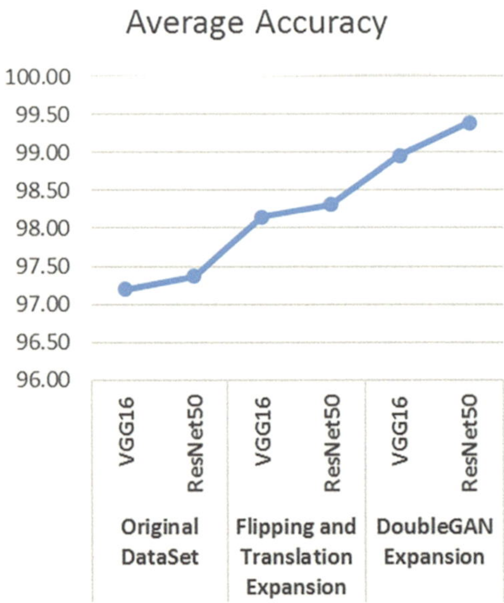


**Figure 5.**
*Plant disease detection model accuracy.*

## 4.3 Crop yield prediction results

In this work, complex ML technologies were employed, such as Boosted Regression Tree (BRT), random forest regression (RFR), support for vector regression (SVR), and Gaussian GPR with grape plants. And these ML techniques are compared with the error parameters MAE, RMSE, and R. The following section and **Figure 6** shows that the Boosted regression tree method has lower error rates than RFR, SVR, and GPR. Finally, based on all these application's outputs, such as irrigation scheduling, plant disease detection, and yield time prediction, the final result can be fine-tuned, i.e., the production of irrigation scheduling can be predicted effectively and efficiently based on plant disease detection output and crop yield time prediction (**Table 3**).

The suggested methodology will be assessed using classification techniques, such as the Decision tree method, random forest method, and naive Bayes

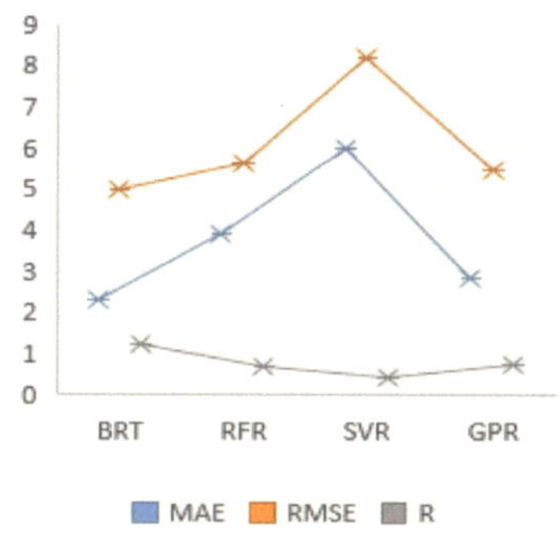


**Figure 6.**
*Yield prediction methods error rate.*

| ML Model | MAE | RMSE | R |
|---|---|---|---|
| BRT | 2.32 | 4.96 | 1.21 |
| RFR | 1.91 | 3.63 | 0.71 |
| SVR | 6.02 | 8.21 | 0.43 |
| GPR | 2.85 | 5.48 | 0.75 |

**Table 3.**
*Yield prediction methods error rate.*

methods. Among these three machine learning models, the random forest yielded superior outcomes. The identification of plant diseases is conducted via the DoubleGAN technique, while the assessment of its performance is carried out using the ResNet and VGG algorithms. Using the DoubleGAN picture dataset, in conjunction with ResNet and VGG models, yields higher accuracy rates than both the original dataset images and conventional enhanced images. The crop yield forecast error rate is calculated using several machine learning algorithms, including BRT, RFR, SVR, and GPR. The support vector regression has a much lower error rate than other approaches.

## 5. Conclusion

The forecast of crop output has great importance in enhancing the pace of production. The dwindling population of seasoned farmers underscores the need to comprehend optimal crop production methods in regions with limited resources, particularly for novice agricultural practitioners. Despite the existence of other studies conducted in this particular domain, the findings of this research indicate that the use of maximum rendering techniques may provide advantageous outcomes due to a multitude of aspects, including the optimization of watering schedules and the facilitation of plant disease diagnosis. A viable methodology for precision agriculture has been devised using N stage Convolutional Neural Networks, which have been integrated with machine-learning techniques. The experimental findings demonstrate that this method provides a pragmatic and accurate outcome by considering irrigation scheduling, disease detection, and crop production prediction. This study has the potential for future expansion by using more GAN approaches that generate a larger quantity of pictures. By doing so, the accuracy of disease identification and crop production prediction may be significantly enhanced.

## Acknowledgements

Dr. T. Karthick, Department of Data Science and Business Systems, SRM Institute of Science and Technology, Chennai, India, supported and co-authored this study. So I thank Dr. T. Karthick for his excellent technical support.

## Conflict of interest

This paper Titled "Precision Agriculture System using Big Data Analytics in Predicting Yield by Controlling Plant Disease and Irrigation Scheduling," by the authors Mr. N. Vasudevan and Dr. T. Karthick declares that the publishing of this paper does not involve any conflicts of interest. The authors certified that the article was plagiarism-free.

## Author details

Vasudevan N.* and Karthick T.
Department of Data Science and Business Systems, School of Computing, SRM Institute of Science and Technology, Chennai, India

*Address all correspondence to: vn8049@srmist.edu.in

## References

[1] Deng T. A survey of convolutional neural networks for image classification: Models and datasets. In: 2022 International Conference on Big Data, Information and Computer Network (BDICN). Sanya, China: IEEE; 2022. pp. 746-749. DOI: 10.1109/BDICN55575.2022.00145

[2] Aishwarya R, Stephanie PAF, Yogitha R, Srinivas CD. Aquatic plant disease detection using deep learning. In: 2023 9th International Conference on Advanced Computing and Communication Systems (ICACCS). Coimbatore, India: IEEE; 2023. pp. 426-430. DOI: 10.1109/ICACCS57279.2023.10113002

[3] Shivaprasad K, Wadhawan A. Deep learning-based plant leaf disease detection. In: 2023 7th International Conference on Intelligent Computing and Control Systems (ICICCS). Madurai, India: IEEE; 2023. pp. 360-365. DOI: 10.1109/ICICCS56967.2023.10142857

[4] Archana U, Khan A, Sudarshanam A, Sathya C, Koshariya AK, Krishnamoorthy R. Plant disease detection using ResNet. In: 2023 International Conference on Inventive Computation Technologies (ICICT). Lalitpur, Nepal: IEEE; 2023. pp. 614-618. DOI: 10.1109/ICICT57646.2023.10133938

[5] Guo J. Research on artificial intelligence: Deep learning to identify plant species. In: 2022 International Conference on Machine Learning and Knowledge Engineering (MLKE). Guilin, China: IEEE; 2022. pp. 59-66. DOI: 10.1109/MLKE55170.2022.00017

[6] Tawegoum R, Leroy F, Sintes G, Chassériaux G. Predictive irrigation scheduling modeling in nurseries. In: Current Perspective on Irrigation and Drainage. London, UK, London, UK: IntechOpen; 2017. DOI: 10.5772/67341

[7] Adeyemi T, Grove I, Peets S, Domun Y, Norton T. Dynamic neural network modelling of soil moisture content for predictive irrigation scheduling. MDPI Sensors. 2018;**18**(10):3408. DOI: 10.3390/s18103408

[8] Mokhtar U, Alit MAS, Hassenian AE, Hefny H. Tomato Leaves Diseases Detection Approach Based on Support Vector Machines. Cairo, Egypt: IEEE; 2015. pp. 246-250

[9] Ali H, Lali MI, Nawaz MZ, Sharif M, Saleem BA. Symptom based automated detection of citrus diseases using color histogram and textural descriptors, Elsevier. Computers and Electronics in Agriculture. 2017;**138**(3):92-104

[10] Zhang S, Xiaowei W, You Z, Zhang L. Leaf image based cucumber disease recognition using sparse representation classification, Elsevier. Computers and Electronics in Agriculture. 2017;**134**:135-141

[11] Jain L, Singh SP, Pritamdas K, Devi KJ, Devi SD. Custom convolutional neural network for detection and classification of rice plant diseases. Procedia Computer Science. 2023;**218**:2026-2040. DOI: 10.1016/j.procs.2023.01.179.22

[12] Zhang X, Qiao Y, Meng F, Fan C, Zhang M. Identification of maize leaf diseases using improved deep convolutional neural networks. IEEE Access. 2018;**6**:30370-30377. DOI: 10.1109/ACCESS.2018.2844405

[13] Chouhan SS, Kaul A, Singh UP, Jain S. Bacterial foraging optimization

based radial basis function neural network (BRBFNN) for identification and classification of plant leaf diseases: An automatic approach towards plant pathology. IEEE Access. 2018;**6**:8852-8863. DOI: 10.1109/ACCESS.2018.2800685

[14] Hang J, Zhang D, Chen P, Zhang J, Wang B. Classification of plant leaf diseases based on improved convolutional neural network. MDPI Sensors. 2019;**19**(19):4161. DOI: 10.3390/s19194161

[15] Mao Z, Chen J, Yang M. Multi-branch Structure for Hierarchical Classification in Plant Disease Recognition. Cham: Springer; 2019. DOI: 10.1007/978-3-030-31726-3_45

[16] Singh UP, Chouhan SS, Jain S, Jain S. Multilayer convolution neural network for the classification of mango leaves infected by anthracnose disease. IEEE Access. 2019;7:43721-43729. DOI: 10.1109/ACCESS.2019.2907383

[17] Khan MA, Lali MIU, Sharif M, Javed K, Aurangzeb K, Haider SI, et al. An optimized method for segmentation and classification of apple diseases based on strong correlation and genetic algorithm based feature selection. IEEE Access. 2019;7:46261-46277. DOI: 10.1109/ACCESS.2019.2908040

[18] Dai Q, Cheng X, Qiao Y, Zhang Y. Crop leaf disease image super-resolution and identification with dual attention and topology fusion generative adversarial network. IEEE Access. 2020;**8**:55724-55735. DOI: 10.1109/ACCESS.2020.2982055

[19] Cap QH, Uga H, Kagiwada S, Iyatomi H. LeafGAN: An effective data augmentation method for practical plant disease diagnosis. IEEE Transaction. 2020;**19**(2):1258-1267. DOI: 10.1109/TASE.2020.3041499

[20] Pham TN, Van Tran L, Dao SVT. Early disease classification of mango leaves using feed-forward neural network and hybrid metaheuristic feature selection. IEEE Access. 2020;**8**:189960-189973. DOI: 10.1109/ACCESS.2020.3031914

[21] Abdu AM, Mokji MM, Sheikh UU. Deep Learning for Plant Disease Identification from Disease Region Images. Cham: Springer; 2020. DOI: 10.1007/978-3-030-66645-3_6

[22] Liu X, Min W, Mei S, Wang L, Jiang S. Plant disease recognition: A large-scale benchmark dataset and a visual region and loss reweighting approach. IEEE Transaction. 2021;**30**:2003-2015. DOI: 10.1109/TIP.2021.3049334

[23] Zhao Y, Chen Z, Gao X, Song W, Xiong Q, Junfeng H, et al. Plant disease detection using generated leaves based on DoubleGAN. IEEE Transactions. 2021;**19**(3):1817-1826. DOI: 10.1109/TCBB.2021.3056683

[24] Paudel D, Boogaard H, de Wit A, Janssen S, Osinga S, Pylianidis C, et al. Machine learning for large-scale crop yield forecasting. Agricultural Systems (Elsevier). 2021;**187**:103016. ISSN 0308-521X. DOI: 10.1016/j.agsy.2020.103016

[25] Sharma A, Jain A, Gupta P, Chowdary V. Machine learning applications for precision agriculture: A comprehensive review. IEEE Access. 2021;**9**:4843-4873. DOI: 10.1109/ACCESS.2020.3048415

[26] Zhang Z, Qiao Y, Guo Y, He D. Deep learning based automatic grape downy mildew detection. Frontiers in Plant Science. 2022;**13**:872107. DOI: 10.3389/fpls.2022.872107

[27] Yinka-Banjo C, Ugot O-A. A review of generative adversarial networks and

its application in cybersecurity. Artificial Intelligence Review. 2020;**53**:1721-1736. DOI: 10.1007/s10462-019-09717-4

[28] Saeed K, Lizhi W. Crop yield prediction using deep neural networks. Frontiers in Plant Science. 2019;**10**:621. Available from: www.frontiersin.org/article/10.3389/fpls.2019.00621. DOI: 10.3389/fpls.2019.00621

[29] Reddy DJ, Kumar MR. Crop yield prediction using machine learning algorithm. In: 2021 5th International Conference on Intelligent Computing and Control Systems (ICICCS). Madurai, India: IEEE; 2021. pp. 1466-1470. DOI: 10.1109/ICICCS51141.2021.9432236

[30] Medar R, Rajpurohit VS, Shweta S. Crop yield prediction using machine learning techniques. In: 2019 IEEE 5th International Conference for Convergence in Technology (I2CT). Bombay, India: IEEE; 2019. pp. 1-5. DOI: 10.1109/I2CT45611.2019.9033611

Chapter 6

# Perspective Chapter: Sustainability of the Extensive Livestock Production – The Role of Precision Agriculture in Mediterranean Montado Ecosystem

*João Serrano, Shakib Shahidian and J. Marques da Silva*


## Abstract

Grassland ecosystems are one of the most common land use types worldwide. The sustainability of extensive livestock production systems requires the optimization of soil management, pasture production and animal grazing. Montado is an agro-silvo-pastoral system characterized by a high complexity as a result of the interactions between climate, soil, pasture, trees and animals. It is in this highly unpredictable context that dynamic management decisions have to be made, for example with respect to soil fertilization or pH correction, grazing or animal supplementation. This study shows, through case studies, the potential of the application of Precision Agriculture (PA) strategies and technologies for monitoring indicators of the Mediterranean Montado multifunctional ecosystem, specific environmental results, namely on soil health and functionality or pasture productivity, quality, or biodiversity, in a holistic and sustainable approach. The use, calibration and validation of soil and pasture proximal sensors, animal GPS (global positioning system) receivers and satellite images opens perspectives to explore new solutions to respond to the new economic and environmental challenges and ensures the resilience of this ecosystem.


**Keywords:** Montado ecosystem, sensors, environmental indicators, soil functionality, pasture biodiversity

## 1. Introduction

Montado is an agro-silvo-pastoral system characteristic of the Mediterranean basin, which covers an estimated area of 850,000 square kilometers, mostly constituted by unfavorable soil and climatic conditions [1]. Its complexity reflects the interactions between climate, soil, pasture, trees, and animals. In Portugal this ecosystem predominates in the Alentejo region (**Figure 1a**). This map presents several Montado experimental fields where studies have been carried out within the scope of

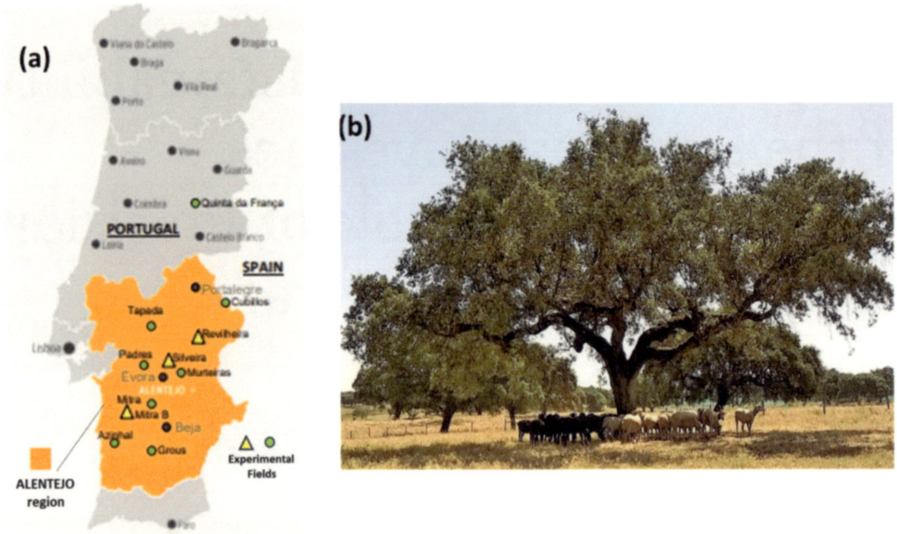


**Figure 1.**
*Location of several Montado experimental fields in the Alentejo region of Portugal (a); illustrative picture of this ecosystem (b).*

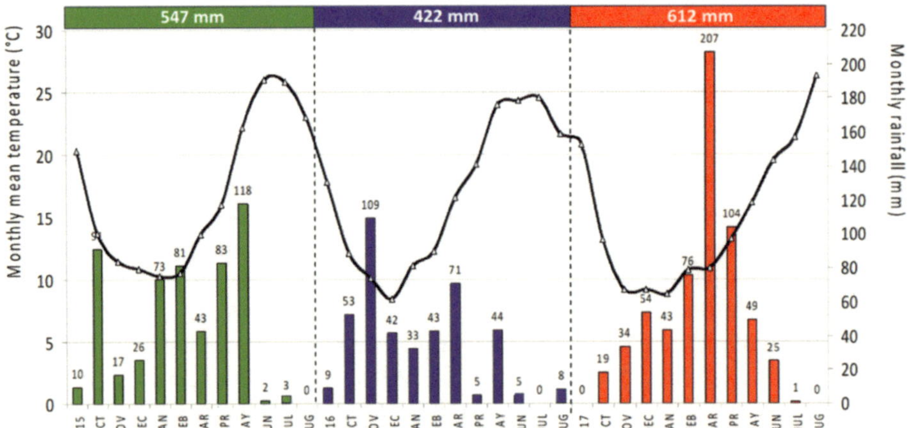


**Figure 2.**
*Monthly mean temperature and monthly rainfall recorded by the meteorological station of "Mitra" (Évora, Portugal), between September 2015 and August 2018.*

research projects to test new monitoring technologies. **Figure 1b** shows a picture of this ecosystem (soil-pasture-trees-animals).

The Alentejo region of Portugal integrates the "Csa-hot-summer Mediterranean climate" of spatial distribution of Köppen-Geiger climate groups developed by AEMET [2], characterized by strong seasonality, with a wide amplitude of temperatures (exceeding 40°C in summer and falling below 0°C in winter), and precipitation (on average between 400 and 600 mm), with strong inter-annual irregularity.

This variability is confirmed in **Figure 2**, which shows temperature and precipitation data recorded at Mitra meteorological station (38°32,2′N; 8°01,1′W; Évora, Alentejo region of Southern Portugal) for three consecutive years (between September 2015 and August 2018) [3]. The accumulated precipitation was 547 mm, 422 mm and 612 mm, respectively, however, while in 2015–2016 the precipitation occurred evenly in the seasons in which there was vegetative development of dryland crops in this region (Autumn, Winter and Spring), 2016–2017 presented a relatively dry spring (120 mm between March and May) and 2017–2018 presented a dry autumn (53 mm between September and November) and a very rainy spring (360 mm between March and May).

Under these conditions pasture availability and quality vary significantly throughout the year, with a surplus of production during spring and scarcity during summer, and from year to year, depending mainly on the distribution of rainfall in the autumn, winter, and spring [4]. **Figure 3** shows the evolution of NDVI obtained by remote sensing (Sentinel-2 satellite imagery) over the period between September 2015 and August 2018, at the "Mitra" experimental field [3]. The standard pattern indicates minimum NDVI values (around 0.2) in the summer months (July-September), when low rainfall is combined with average temperatures around 25°C, and maximum temperatures may reach at 40°C. During autumn (October-December) there is a very significant increase in NDVI, depending on the distribution of rainfall over this period. In winter, usually between February and March, NDVI reaches its peak (around 0.8). From this point on, the clear rise in temperatures accelerates the pasture vegetative cycle, with the NDVI decreasing sharply, depending once again on the amount of rainfall that occurs during this period. In addition to climatic seasonality, during the study period there was an important inter-annual variability. It is in this highly unpredictable context that management decisions have to be made, for example with respect to soil fertilization, grazing or animal supplementation [5].

Recent studies show the potential of the Montado to provide a range of ecosystem services [6]. Guimarães et al. [7] go further and propose a payment model for

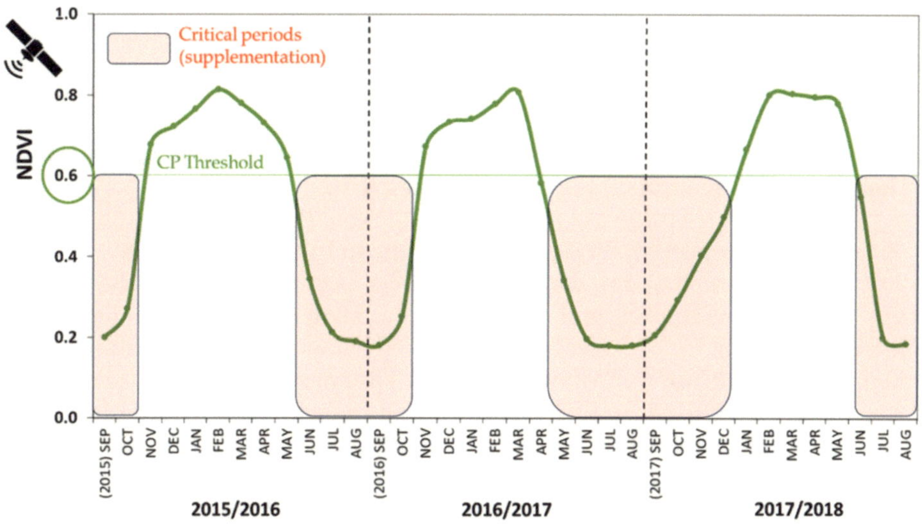


**Figure 3.**
*Evolution of mean normalized difference vegetation index (NDVI) at Mitra experimental field between November 2015 and June 2018. The critical periods of supplementation (NDVI below 0.6, corresponding to sheep maintenance requirements in terms of crude protein, CP; [5]) are indicated.*

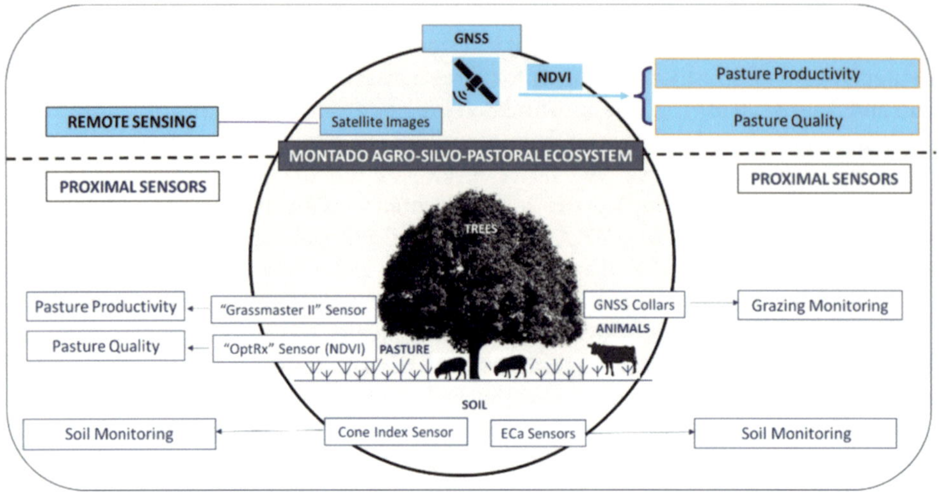


**Figure 4.**
*Technological approach for a holistic view of the Montado ecosystem.*

farmers, implemented under Common Agricultural Policy and based on specific environmental results, namely, a healthy and functional soil ecosystem, a biodiverse native Mediterranean pasture, tree regeneration and conservation of unique elements in the landscape. It is in this framework that expedient technologies such as soil electrical conductivity meter, electronic cone penetrometer, optical sensors, GNSS (Global Navigation Satellite Systems) receivers, satellite imagery and others become indispensable expedient monitoring tools to implement differentiated management strategies and to quantify relevant indicators and that allow the follow-up by public authorities and policy makers [6].

This study aims, through case studies, to demonstrate, the potential of the application of Precision Agriculture (PA) strategies and technologies to monitoring the impact of soil, pasture, tree and animal grazing interactions on soil health and functionality and pasture biodiversity indicators (**Figure 4**).

## 2. From intensive sampling grids to the definition of management zones: the role of soil electrical conductivity sensors

One of the key aspects in PA is the survey of the soil and/or crop spatial variability. Intensive sampling, based on grids, is a reliable process. **Figure 5** shows some examples of regular sampling grids installed by our team in experimental fields in the Montado ecosystem: (a) "Revilheira" (76 sampling points in 7 ha); (b) "Silveira" (47 sampling points in 6 ha); (c) "Mitra Eco-SPAA" (48 sampling points in 4 ha). This procedure, however, is extremely time demanding and with significant costs, making it impractical on a real scale for the farmer. An alternative with great potential is to carry out a survey of the apparent soil electrical conductivity (ECa). Non-contact sensors, electromagnetic induction (e.g., Dualem 1S or EM38; **Figure 6a** and **b**, respectively) or contact sensors (e.g., Veris 2000 XA; **Figure 6c**), associated with GNSS receivers allow the collection of the spatial variability pattern of some soil properties and, through geostatistical algorithms (e.g., Rasch model; [8]) to define management zones (MZ). **Figure 7** shows the ECa map and the indication of two

**Figure 5.**
*Regular sampling grids installed in three experimental farms ("Revilheira"—(a); "Silveira"—(b); "Mitra Eco-SPAA"—(c).*

**Figure 6.**
*Apparent soil electrical conductivity sensors used in the experimental works: "Dualem 1S" (a); "EM-38" (b); and "Veris 2000 XA" (c).*

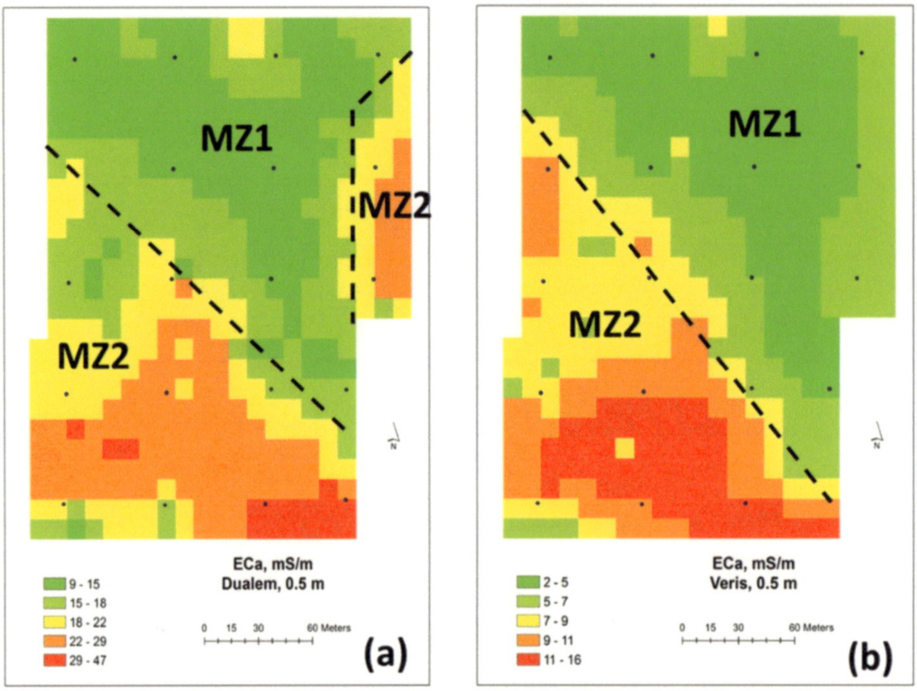


**Figure 7.**
*Apparent soil electrical conductivity maps of "Revilheira" experimental field: obtained from the survey carried out by "Dualem 2S" sensor (a); or by "Veris 2000 XA" sensor (b).*

potential MZ for the purpose of soil smart sampling in the experimental field of "Revilheira" (38°27,9′N; 7°25,8′W). One of the maps (**Figure 7a**) was obtained with a "Dualem 1S" sensor and the other (**Figure 7b**) was obtained from the survey carried out with a "Veris 1S" sensor, showing very similar patterns of variability. Based on soil sampling by zones, differentiated application of fertilizers or amendments can be implemented using variable rate application (VRA) technology.

## 3. Grazing monitoring with GNSS collars: evaluation of the impact of animal trampling on soil compaction

Monitoring grazing is a tool to support livestock management, with the aim of optimizing available resources, especially in terms of feed, reproduction, health, and animal welfare. The traditional way is carried out by experienced observers and is based on regular and "in loco" observation and monitoring of the herd, seeking to identify and record revealing behavior patterns. GNSS technology has the costs associated with the acquisition of collars and the respective licenses, but it provides remote monitoring of the animals, which can result in important benefits. The use of GNSS collars on animals introduced an important technological advancement, allowing detailed and practically continuous information to be obtained on grazing pathways, preferred grazing zones and areas of higher stocking densities. Consequently, better knowledge can be obtained about the interactions between several components of the Montado ecosystem, which is essential as a support tool for making decisions [9].

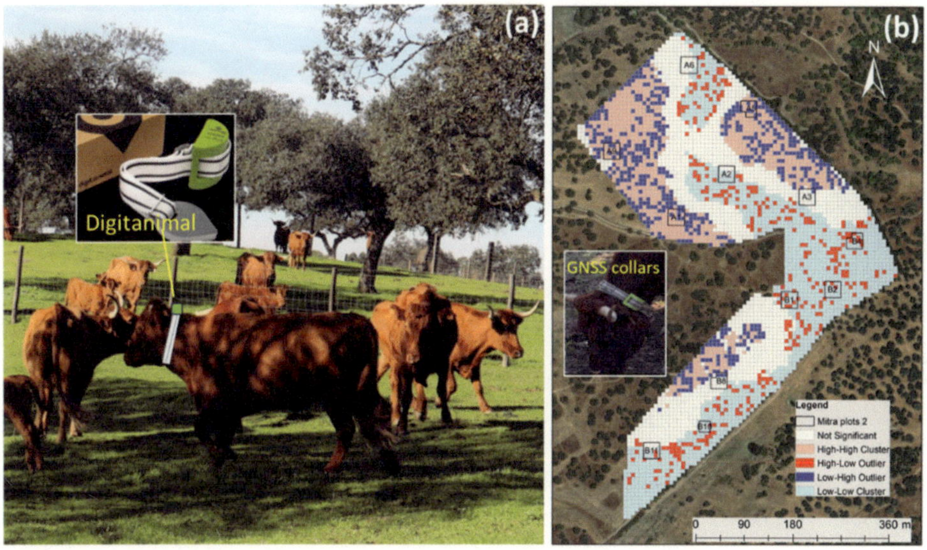


**Figure 8.**
*(a) Cows grazing at "Mitra B" farm (detail of GNSS receivers); (b) grazing density map based on georeferenced information obtained by GNSS collars.*

Our research team used GNSS position loggers (CatTrackTM, Perthold Engineering LLC, Anderson, USA) on sheep in the experimental field "Mitra-Eco-SPAA" for 22 weeks (from 1 April to 25 August 2016), which allowed for the monitoring of seasonal changes of feeding sites selected by grazing ewes [9].

Others GNSS position loggers ("Digitanimal", Madrid, Spain; **Figure 8a**) were used in cows in the experimental field "Mitra B" between January and May 2021, allowing for the monitoring of grazing patterns. Geostatistical analyses of the GNSS collars were carried out with the ArcGIS Desktop software (v10.5, ESRI, Inc., Redlands, CA, USA). The "Optimized Outlier Analysis (Spatial Statistics)" algorithm, based on incident points, obtained by the GNSS collars, creates a map of statistically significant hot spots, cold spots and spatial outliers using the "Anselin Local Moran's I statistic". This map, with a 5 m spatial resolution, includes 5 classes ("Not significant", "High-High cluster", "High-Low outlier", "Low-High outlier", and "Low-Low cluster") and serves to characterize the grazing density pattern (**Figure 8b**). This work allowed for the evaluation of the effect of livestock trampling on soil compaction of "Mitra-B" experimental field. For this purpose, an electronic cone penetrometer "FieldScout SC 900" (Spectrum Technologies, Aurora, IL, USA; **Figure 9a**) was used to measure the soil resistance to penetration (cone index, CI, in kPa) in sampling areas and in three dates (03 November 2021, 21 December 2021 and 15 March 2022; Figures b, c and d, respectively). For the statistical analysis, only two classes were considered: the "High-High cluster" (a statistically significant cluster of high values), and the "Low-Low cluster" (a statistically significant cluster of low values). Our results show a systematic trend of a greater CI in areas with "high" livestock trampling, compared to areas with "low" livestock trampling [10].

Soil compaction is generally viewed as an indicator of soil degradation processes and a determinant factor in crop productivity. In the Montado silvo-pastoral ecosystem, animal trampling is mentioned as a variable to consider in soil compaction. To mitigate the negative effects of livestock trampling, according to Reichert et al. [11],

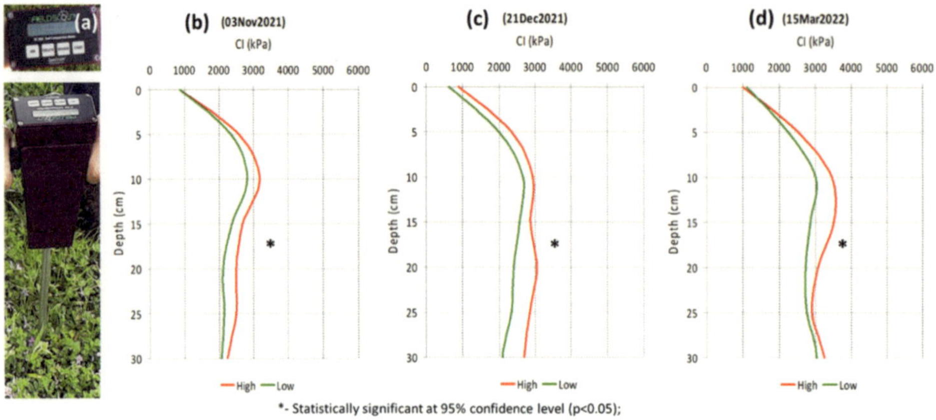


**Figure 9.**
*Electronic cone penetrometer "FieldScout SC 900" (a); average cone index (CI) at 0–0.30 m soil depth, in high and low-grazing-intensity areas: (b) 3 November 2021; (c) 21 December 2021; (d) 15 March 2022.*

two preventative measures may be implemented: (i) the use of mobile fences to impede the entry of animals into more susceptible areas in the days following the occurrence of major precipitation events; (ii) control over grazing management (stocking rates) to permanently provide a minimum height of pasture to ensure soil surface protection. One of the main causes of soil and pasture degradation resulting from livestock trampling is the natural tendency of animals to agglomerate in certain areas of the pasture. Thus, an understanding of the potential environmental effects of the concentration areas is necessary to adequately consider mitigating grazing management practices [12]. Some examples of grazing management practices that most influence grazing distribution and, consequently, livestock trampling, are the location of watering sites, supplement feeders, tree shade or fence and gate placement [12]. Compaction at gates and waterers, for example, is inevitable, but can be minimized. The traditional advice of moving waterers and feeders, fences, and gates to avoid systematic and preferential routes and to limit livestock concentration areas seems appropriate [12]. Temporary livestock exclusion areas (buffer zones) can be an attractive tool to improve soil physical properties in susceptible areas.

One of the open questions associated with animal trampling and the impact on soil compaction is related to grazing systems [6]. Mediterranean pasture ecosystems are usually extensive. Therefore, in this region the predominant system is the continuous grazing (CG), which is a grazing management modality with a permanent and low animal stocking rate (≤1 unit animal, UA, per hectare, corresponding to approximately 6–7 sheep $ha^{-1}$) [6] and without a grazing control based on the amount of effective forage accumulation (height monitoring) [6]. It is known that increased stocking rates have negative effects on soil properties and are positively correlated with animal trampling. Donkor et al. [13], for example, suggested that deferred grazing (DG) systems have associated greater CI than CG systems. This was the starting point for a study conducted by our team, which involved measurements with the electronic cone penetrometer "FieldScout SC 900" and aimed at assessing the impact of sheep deferred grazing (DG; that consists in adapting the number of animals and the number of days grazed to the availability of pasture; in this study stocking rate was of 2 UA) on soil compaction as a result of animal trampling. This study was carried out between September 2019 and June 2021, in "Mitra Eco-SPAA" experimental farm.

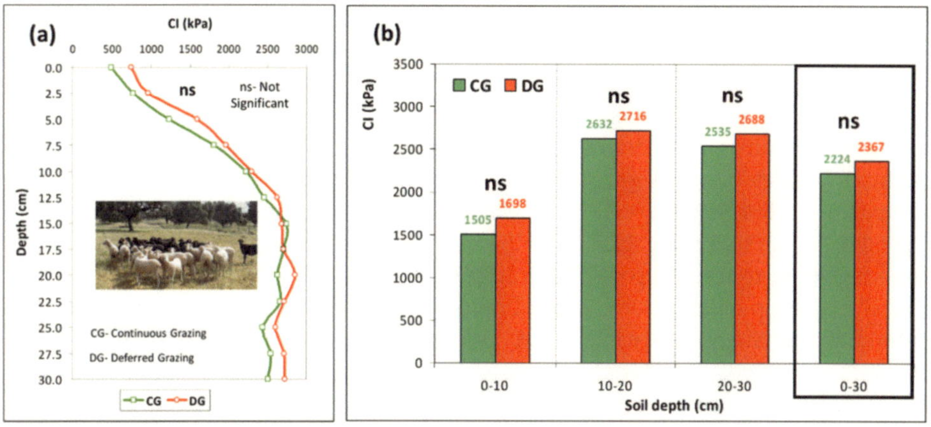


**Figure 10.**
*Average cone index (CI, in kPa) for continuous (CG) versus deferred grazing (DG) at: (a) 0–30 cm soil depth; (b) 0–10 cm, 10–20 cm, 20–30 cm and 0–30 cm soil depths.*

The results (CI, in kPa) are presented in **Figure 10**. Although average CI values tend to be higher in areas with DG (**Figure 10a**) at all depths (**Figure 10b**), the differences obtained are not statistically significant ("ns"; p = 0.337). However, this systematic tendency suggests that future works of longer duration to evaluate compaction/recovery cycles, should also consider the trend towards increased intensification of extensive livestock production, based on electric and mobile fences, with a stocking rates of up to three times higher than in CG systems (approximately 1 UA, per hectare, corresponding to approximately 6–7 sheep $ha^{-1}$). DG can be used as a tool to improve pasture resilience, livestock performance, pasture quality and profitability at a farm scale, when considering a typical spring surplus [14, 15]. Dynamic grazing systems, that combine DG with flexible stocking rate based on changing rainfall conditions, are fundamental for achieving sustainable outcomes [14, 16]. The prospect of intensifying extensive forage-based livestock through the adoption of dynamic grazing systems should take into consideration the long-term impact on soil compaction and, mainly, the system's productivity indicators.

Nevertheless, long term studies are required [17] to quantify responses of different pasture types under variable climatic scenarios and, mainly, whole-farm analysis to integrate the multiple impacts of DG on the farm system, which include the positive impacts on the pasture performance after the deferred period, and the negative impact on pasture growth and nutritive value during the deferred period [16]. These studies should also include the measurement of parameters indicating change in ecosystem function, resilience, and ecological services [14, 15]. Recent studies show the potential of the Montado to provide a range of ecosystem services. Guimarães et al. [7] go further and propose a results-based model implemented under the Common Agricultural Policy and based on specific environmental results, namely, a healthy and functional soil ecosystem and a biodiverse native Mediterranean pasture. For example, soil degradation is identified as an important problem in the Montado, where animal grazing activity impacts soil health and its productivity through trampling [7]. It is in this framework that expedient technologies such as the electronic cone penetrometer, the soil electrical conductivity meter, satellite imagery and others will become indispensable monitoring tools for the quantification of subsidies to farmers and for the follow-up by public authorities and policy makers.

## 4. Pasture monitoring: proximal and remote sensing

### 4.1 Pasture productivity monitoring

The seasonality and interannual irregularity of pasture availability (productivity and quality) is evident in **Figures 2** and **3** (respectively, the patterns of temperature/precipitation, and NDVI). The pasture availability ensures the animal feed in extensive production systems. The decision-making of animal grazing management (stocking rates, rotation through the various fields, calculation of forage needs, or animal feed supplementation) is based on pasture productivity [18]. The standard method of assessing pasture mass (harvesting green matter at specific sampling areas), is a lengthy and expensive process given the large number of samples necessary to accurately represent pasture spatial variability [19] and farmers cannot make this effort in day-to-day management of pastures [18]. Therefore, other techniques, which do not harm the pasture and are useful over large areas, are required to clarify yield variability and implement more contemporary production strategies such as precision agriculture or site-specific management.

A long-term study (2007–2019) to calibrate a Grassmaster II capacitance probe (introduced by Novel Ways Electronic, a New Zealand company; **Figure 11**) to estimate pasture productivity was carried out in Alentejo region of Portugal [18, 20].

The working principle of this probe is as follows: an electronic circuit generates a signal of a certain frequency, and then carries out a measurement of the capacitance of the air-herbage mixture. The probe produces an electric field that is modified by the pasture close to it. The capacitance meter senses an area approximately 100 mm in diameter and 400 mm in length. The modified field is detected as a change in capacitance by the electronic circuit within the probe. This capacitance change is proportional to the water content of the grass. Because the dry matter content is

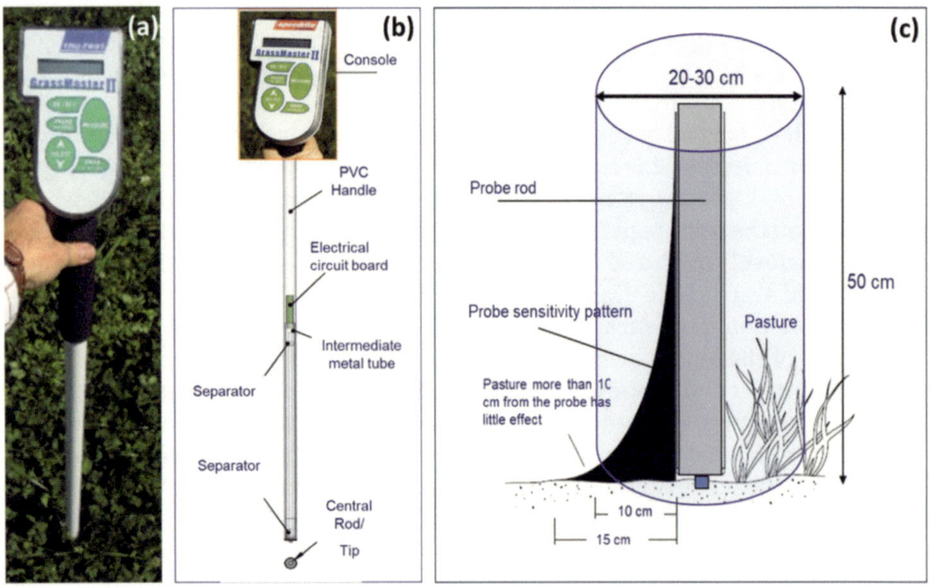


**Figure 11.**
*Grassmaster II probe in the field (a); constitution (b) and probe sensitivity pattern (c).*

| PMC classes, % | n | Linear equations | r | RMSE, kg ha$^{-1}$ | $CV_{RMSE}$, % |
|---|---|---|---|---|---|
| >80 | 567 | GM = 3.2963*CMR – 9392.7 | 0.775** | 4806 | 28.1 |
| 70–80 | 418 | GM = 1.9042*CMR – 1449.2 | 0.572* | 6839 | 63.0 |
| <70 | 426 | GM = 1.5881*CMR – 411.24 | 0.470* | 4837 | 60.9 |
| >80 | 567 | DM = 0.4656*CMR – 1174.8 | 0.750** | 763 | 29.7 |
| 70–80 | 418 | DM = 0.4483*CMR – 260.57 | 0.567* | 1543 | 58.5 |
| <70 | 426 | DM = 0.4222*CMR + 953.26 | 0.298 ns | 2122 | 66.8 |

*PMC—pasture moisture content; n—number of samples; CMR—"Grassmaster II" measurements; GM—green matter; DM—dry matter; r—correlation coefficient; RMSE—root mean square error; $CV_{RMSE}$—coefficient of variation of root mean square error; ns—not significant.* *Correlation significant at the 0.05 level.*
***Correlation significant at the 0.01 level.*

**Table 1.**
*Proposed linear regression equations for estimating pasture green matter yield (GM, in kg ha$^{-1}$) and dry matter yield (DM in kg ha$^{-1}$) based on "Grassmaster II" measurements (CMR), for each pasture moisture content (PMC) class considered.*

highly correlated with the water content, the probe capacitance change is calibrated to the pasture yield, with linear calibration equations used to correlate readings taken on pasture over a short seasonal time, with measured dry matter production [20].

The relation between pasture productivity and capacitance measured by the probe (CMR) is influenced by factors such as the mixture of plant species and their phenological stage, and their consequent effect on pasture moisture content (PMC) [21]. **Table 1** shows the linear regression equations between green matter yield (GM) and CMR and between dry matter yield (DM) and CMR, for each PMC class considered (>80%; 70–80% and < 70%). For each regression equation, the correlation coefficient (r) and the root mean error (RMSE) are presented in absolute value and in percentage ($CV_{RMSE}$). The highest correlation coefficients and the lowest $CV_{RMSE}$ (better predictability) occur for PMC >80%. As PMC decreases, these indicators deteriorate: the correlation coefficient decreases and the $CV_{RMSE}$ increases. As might

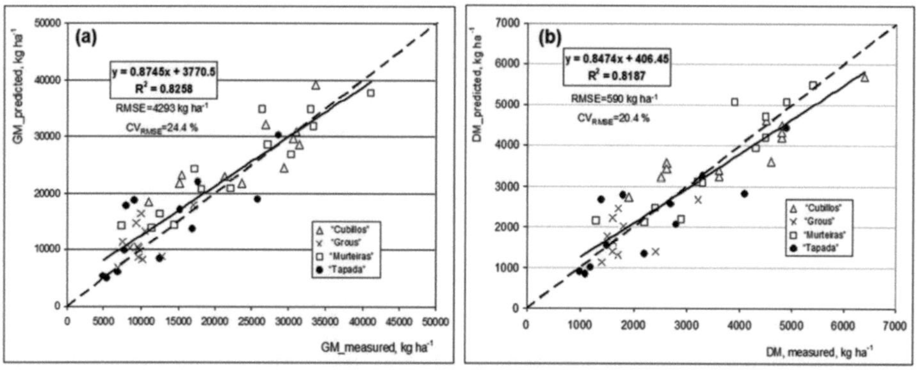


**Figure 12.**
*Relation between measured pasture productivity ((a) green matter; (b) dry matter (DM)) in four experimental fields of Alentejo region ("Cubillos"; "Grous"; "Murteiras" and "Tapada") and predicted pasture productivity based on the equations obtained in "Grassmaster II" measurements and calibration tests.*

be expected in view of the operating principle of the capacitance probe (sensitivity to PMC), water is, by far, the material that affects the probe signal the most [21].

This variability requires the calibration tests to span the diversity of parameters related to permanent dry land pastures of Alentejo area of Portugal. **Figure 12** shows the relation between pasture productivity measured in 2019 in four experimental fields of Alentejo region ("Cubillos"; "Grous"; "Murteiras" and "Tapada"; **Figure 1**) and pasture productivity prediction based on the general equation obtained for PMC > 80% (**Table 1**), proving a very acceptable degree of confidence ($CV_{RMSE}$ of 24.4% and 20.4%, respectively to GM and DM) [21]. These results are very promising, especially given the heterogeneity of biodiverse pastures. Each pasture is a different ecosystem, with specific characteristics, which vary according to the different plant species and their vegetative states. Deviations of approximately 20–25% between the measured and the estimated productivity are, in practice, not impeditive to support precision management decisions in grazing systems [21]. It can be concluded that the capacitance probe is an expedient tool to support the decision-making process in the management of dry land pasture and its respective dynamic grazing.

### 4.2 Pasture quality monitoring

Pasture quality can be quantified by certain parameters, such as crude protein (CP) and neutral detergent fiber (NDF) content [22]. It is known that high CP content and low NDF content indicate higher quality pastures [23], which is usually the case when the plants are young. For this reason, PMC can also be considered a parameter of pasture quality [24]. As the pasture develops during the spring, PMC and CP tend to decrease while the NDF tends to increase [25]. The conventional method for obtaining these indicators of pasture quality (PMC, CP and NDF) is based on in-situ sampling and chemical analysis. However, this standard methodology involves intensive field work and lengthy laboratory analysis, which are relatively expensive and sometimes belated, making them unpractical for day-to-day farm management [26].

In recent years, several works have been developed that show the application of new technologies and methodologies for the estimation of pasture quality. Some of the more successful technologies are based on vegetation indices (VI) obtained from the measurement of the reflectance of the radiation received by the plants [27]. VI can help assess crop growth, vigor, biomass, or chlorophyll content and, consequently, help understand the patterns of spatio-temporal variability, which is crucial for PA applications [27]. The Normalized Difference Vegetation Index (NDVI) is the most common vegetation assessment because of its ability to quickly delineate vegetation

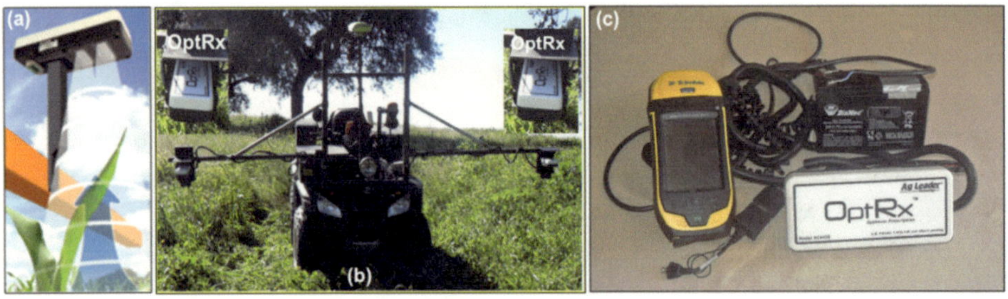


**Figure 13.**
*Proximal active optical sensor ("OptRX™") used to measure NDVI: (a) plant reflectance; (b) pasture field monitoring; (c) sensor, small battery and GNSS receiver.*

and vegetative stress [28]. The mathematical expression that allows the calculation of NDVI considers the reflectance of vegetal tissues in the red and near-infrared (NIR) ranges of the electromagnetic spectrum (Eq. (1)), used to estimate, among other parameters, the plants chlorophyll content [29]. Therefore, the solution for quick monitoring pasture productivity may be using proximal sensors (PS) and non-destructive techniques [30]. Active optical sensors (AOS) when mounted on mobile platforms, combined with the use of global navigation satellite systems (GNSS), have the potential to deliver continual data on the feed base, to capture rapid changes in the vegetation reflectance response and, consequently, in the proportions of photosynthetically active vegetation [30].

$$NDVI_{PS} = \frac{NIR - RED}{NIR + RED} \tag{1}$$

In our study [23, 25] a commercial "OptRx™" AOS (**Figure 13**), manufactured by Ag Leader (Ames, IA, USA), was used for vegetation multispectral measurements. This sensor was associated with a Trimble GNSS GeoExplorer 6000 series model 88,951 (Trimble Navigation Limited, Raunheim, Germany), with sub-meter precision. The sensor was transported by a mobile platform at an average speed of 5 km $h^{-1}$.

Other methods proposed for quantifying and mapping the pasture quality variability are based on NDVI calculated by measuring the reflectance of the radiation emitted by the plants, using satellite images [31, 32]. In our study [23], Sentinel-2 band 4 (B4; 10 m spatial resolution; 665 nm) and band 8 (B8; 10 m spatial resolution; 842 nm), atmospherically corrected imagery, were extracted from Copernicus data hub and used to calculate $NDVI_{RS}$ (Eq. (2)).

$$NDVI_{RS} = \frac{B8 - B4}{B8 + B4} \tag{2}$$

**Figure 14** shows an example of Sentinel-2 NDVI pixels (spatial resolution of 10 m) obtained from our experimental fields. In this experimental field (Mitra "Eco-SPAA"), with 4 ha, one hundred and seventy NDVI pixels without tree effect were established. To conduct the reconstruction of historical NDVI trends, NDVI time series of these pixels were extracted throughout the pasture vegetative cycle.

**Figure 15** shows an example of NDVI maps obtained by Sentinel-2 on three dates: November 2020 (a), April 2021 (b) and July 2021 (c). Beyond the spatial variability, these maps show the temporal evolution of the pattern of pasture vegetative vigor, from emergence (October/November) to peak production (April/May), also reflecting the break in late spring (June) and summer (July/September) [33].

Although use of satellite imagery is a very promising, low cost and non-destructive technique, it has its own limitations in terms of spatial resolution and the presence of clouds, as well as the spatial and temporal specificity of the associated algorithms [30]. Therefore, a viable alternative to overcome these difficulties could be the fusion of the RS data with PS [34], by extracting complementary information from multiple sensors or sources, simultaneously solving the difficulty associated with access to under the tree canopy in the Montado ecosystem [25]. In a study carried out by our team to assess the potential of two complementary tools (satellite images, Sentinel-2 and proximal optical sensor, OptRx) for the calculation of NDVI and, thus, to obtain in a timely manner, indicators of pasture quality [24], in two consecutive years (2018/2019 and 2019/2020), six fields representative of dryland pastures in the

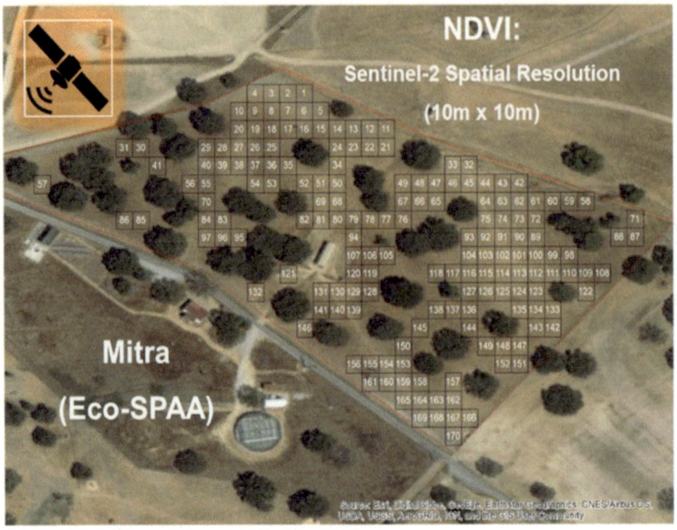


**Figure 14.**
*Sentinel-2 pixels (10 m × 10 m) used in Mitra "Eco-SPA" field.*

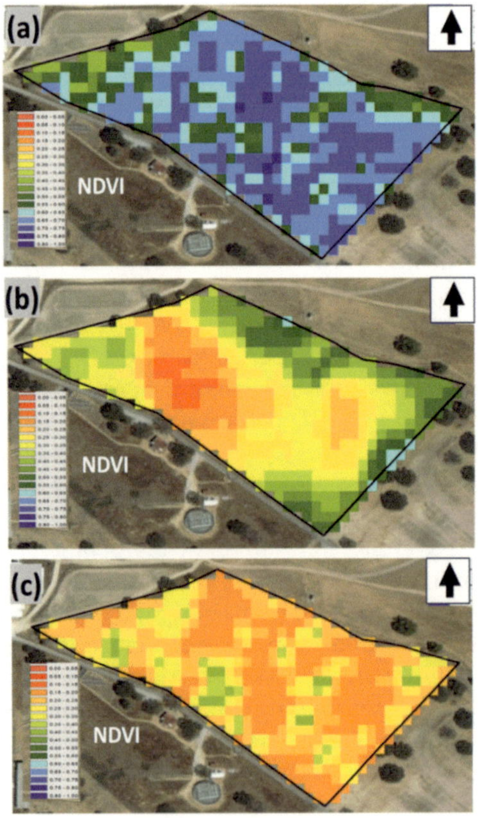


**Figure 15.**
*Normalized difference vegetation index (NDVI) maps of Mitra "Eco-SPAA" field obtained by Sentinel-2 on three dates: November 2020 (a), April 2021 (b) and July 2021 (c).*

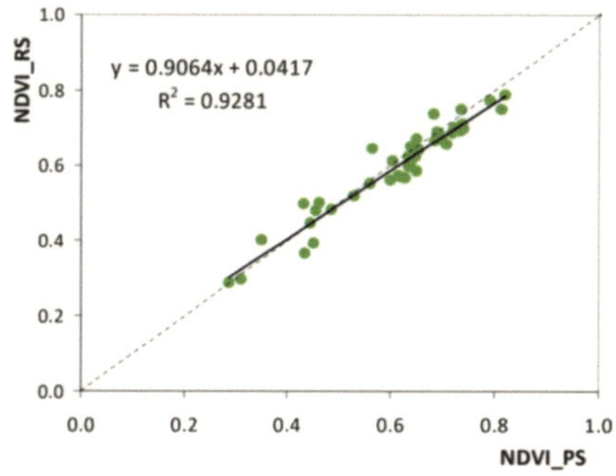


**Figure 16.**
*Regression analysis between vegetation index values obtained by proximal sensor, $NDVI_{PS}$ and by satellite, $NDVI_{RS}$.*

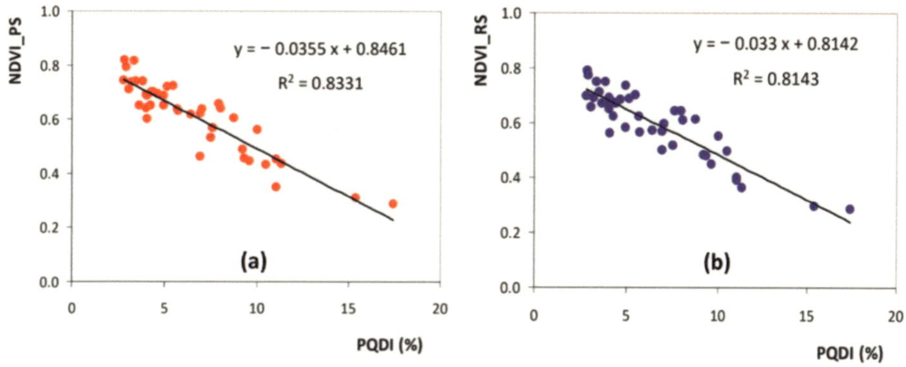


**Figure 17.**
*Regression analysis between: (a) vegetation index values obtained by proximal sensor, $NDVI_{PS}$ and PQDI; (b) vegetation index values obtained by satellite, $NDVI_{RS}$ and PQDI.*

Alentejo region of Portugal were monitored. To simplify the approach, an index of pasture quality degradation (PQDI) has been calculated that simultaneously integrates PMC, CP and NDF (Eq. (3)).

$$PQDI = \frac{NDF}{PMC \times CP} \times 100 \quad (3)$$

The results of the regression analysis show a strong coefficient of determination between NDVI obtained by RS and NDVI obtained by PS ($R^2$ = 0.93; **Figure 16**), and between any of these and PQDI: $R^2$ = 0.83 between PQDI and $NDVI_{PS}$ (**Figure 17(a)**); $R^2$ = 0.81 between PQDI and $NDVI_{RS}$ (**Figure 17(b)**). Therefore, it can be inferred that these technological tools (PS and RS) can potentially make an important contribution to decision making and to the management of livestock production.

## 5. Indicative cost and potential of this technological approach in Montado ecosystem

From a scientific point of view, all the sensors used and trialed in this study are highly valuable in terms of producing knowledge for the operational and differential

| Proximal sensor | Parameter measured | Cost per unit, € | Additional cost, € |
|---|---|---|---|
| *Soil sensors* | | | |
| FieldScout SC 900 | CI, kPa | 3700 | 3500* |
| Veris 2000 XA or Dualem 1S | ECa, mS $m^{-1}$ | 25,000–30,000 | Not req. |
| *Pasture sensors* | | | |
| Grassmaster Pro | DM or GM, kg $ha^{-1}$ | 2700 | 3500* |
| OptRx | NDVI | 4250 | 3500* |
| *Animal sensor* | | | |
| GPS collars | Coordinates | 200 | 2000** |

*GPS—global positioning system; CI—cone index; ECa—apparent soil electrical conductivity; DM—dry matter; GM—green matter; NDVI—normalized difference vegetation index; Not req.—not required. *Georeferencing System (GPS unit).*
***Annual license.*

**Table 2.**
*Indicative cost of soil, pasture and animal proximal sensors used in this study.*

management of the Montado ecosystem [5]. For reference purposes, the commercial sensors used in this study can be found on the market at the prices shown in **Table 2**, including taxes and transport costs.

These high initial costs reinforce the difficulty of acquiring this technology by the farmer and highlight the interest and potential for service providers, who can benefit from the scale factor. For example, it is possible to access ECa survey services at around 50 € $ha^{-1}$, already including ECa mapping and decision support (with fertilizer prescription maps for variable rate technology, VRT). On the other hand, from the point of view of its applicability by the farm manager, remote sensing based on satellite imagery ("Sentinel-2") seems to be the most suitable, due to its low cost and intrinsic characteristics [5]. Here too, service providers offer farmers support ranging from 250 euros to 5000 € $year^{-1}$, whether they only provide NDVI maps or if they also make recommendations.

In terms of creation of scientific knowledge applied to the management of Montado ecosystem, the great challenge during the next years will be to use the more expensive technologies (soil electric conductivity sensors, proximal multispectral sensors, capacitance probes, animal GNSS receivers, etc.) to model and validate methods based on remote sensing. These should be able to monitor, detect, forecast and alert about the Montado ecosystem, ensuring its balanced, holistic and sustainable management [5].

## 6. Conclusions

The current scenario of climate change and its impact on dryland Mediterranean pastures poses new challenges for extensive livestock production and for ecoservices related to this ecosystem. In this context, several technologies are available that allow for the mapping the soil and pasture spatial variability, and pasture vegetative cycle over time and providing real-time information about pasture productivity and quality. The knowledge generated by these sources of information represent a fundamental milestone in supporting decision-making and promoting efficient grazing management.

The results of this study show that data based on temporally stable $EC_a$ survey can be used to define MZ and implement site-specific management of dryland permanent pastures. This is a rational way to improve the efficiency of the use of inputs by adjusting them to soil and pasture variability. The limits of these MZ may be dynamic, allowing the farm manager, in the following years, to make adjustments based on new accumulated knowledge, obtained, for example, by soil and pasture smart sampling and/or a time series of vegetation indices obtained by proximal or remote sensing. Obtaining these indices with interesting spatial and temporal resolution and without cost provide decisive elements for decision-making in the grazing management (e.g., stocking rates, time, and areas of grazing, or need of supplementation).

This exploratory work in the Alentejo region of southern Portugal can be the starting point for the large-scale implementation of this concept, which requires further medium- and long-term validation studies, extended to the other Mediterranean areas, both in terms of cost-benefit analysis (economic and environmental), and impact on livestock system productivity, as well as on the ecosystem biodiversity.

## Acknowledgements

This work was funded by National Funds through the FCT (Foundation for Science and Technology) under the Project UIDB/05183/2020, and by Évora University through of the "Contracts for the Provision of Teaching Services" numbers 7838 and 8179.

## Author details

João Serrano*, Shakib Shahidian and J. Marques da Silva
MED—Mediterranean Institute for Agriculture, Environment and Development and CHANGE—Global Change and Sustainability Institute, Institute of Research and Advanced Training, University of Évora, Pólo da Mitra, Évora, Portugal

*Address all correspondence to: jmrs@uevora.pt

## References


[1] Psyllos G, Hadjigeorgiou I, Dimitrakopoulos PG, Kizos T. Grazing land productivity, floral diversity, and management in a semi-arid Mediterranean landscape. Sustainability. 2022;**14**:4623

[2] FAO. World Reference Base for Soil Resources. Food and Agriculture Organization of the United Nations, World Soil Resources Reports N 103. Rome, Italy: FAO; 2006

[3] Serrano J, Shahidian S, da Silva JM, Paixão L, Carreira E, Pereira A, et al. Climate changes challenges to the management of Mediterranean Montado ecosystem: Perspectives for use of precision agriculture technologies. Agronomy. 2020;**10**:218

[4] Efe Serrano J. Pastures in Alentejo: Technical Basis for Characterization, Grazing and Improvement. Évora, Portugal: University of Évora—ICAM; 2006. pp. 165-178

[5] Serrano J, Shahidian S, Marques da Silva J, Carvalho M. A holistic approach to the evaluation of the Montado ecosystem using proximal sensors. Sensors. 2018;**18**:570

[6] Serrano J, Carreira E, Shahidian S, Carvalho M, Marques da Silva J, Paniagua LL, et al. Impact of deferred versus continuous sheep grazing on soil compaction in the Mediterranean Montado ecosystem. Agriengineering. 2023;**5**:761-776

[7] Guimarães MH, Pinto-Correia T, Costa Freitas MB, Ferraz-de-Oliveira I, Sales-Baptista E, Ferragolo da Veiga JF, et al. Farming for nature in the Montado: The application of ecosystem services in a results-based model. Ecosystem Services. 2023;**61**:101524

[8] Moral F, Terrón J, da Silva JM. Delineation of management zones using mobile measurements of soil apparent electrical conductivity and multivariate geostatistical techniques. Soil & Tillage Research. 2010;**106**:335-343

[9] Serrano J, Sales-Baptista E, Shahidian S, Marques da Silva J, Ferraz de Oliveira I, Lopes de Castro J, et al. Proximal sensors for monitoring seasonal changes of feeding sites selected by grazing. Agroforestry Systems. 2021;**95**:55-69

[10] Serrano J, Marques J, Shahidian S, Carreira E, Marques da Silva J, Paixão L, et al. Sensing and mapping the effects of cow trampling on the soil compaction of the montado Mediterranean ecosystem. Sensors. 2023;**23**:888

[11] Reichert JM, Suzuki LE, Reinert DJ. Compactação do solo em sistemas agropecuários e florestais: Identificação, efeitos, limites críticos e mitigação. Tópicos em Ciências do Solo. 2007;**5**:49-134. (In Portuguese)

[12] Sanderson MA, Feldmann C, Schmidt J, Herrmann A, Taube F. Spatial distribution of livestock concentration areas and soil nutrients in pastures. Journal Soil Water Conservation. 2010;**65**:180-189

[13] Donkor NT, Gedir JV, Hudson RJ, Bork EW, Chanasyk DS, Naeth MA. Impacts of grazing systems on soil compaction and pasture production in Alberta. Canadian Journal of Soil Science. 2002;**82**:1-8

[14] Ma L, Yuan F, Liang H, Rong Y. The effects of grazing management strategies on the vegetation, diet quality, intake and performance of free grazing sheep. Livestock Science. 2014;**161**:185-192

[15] Wang T, Teague WR, Park SC. Evaluation of continuous and multipaddock grazing on vegetation and livestock performance—A modelling approach. Rangeland Ecology & Management. 2016;**69**:457-464

[16] Tozer KN, Muller K, Craven T, Tarbotton I, Coster A, Burke R, et al. Effect of deferred grazing during late spring and summer on pasture productivity in Waikato and Bay of Plenty hill country. Journal of New Zealand Grasslands. 2020;**82**:111-119

[17] Glindemann T, Wang C, Tas BM, Schiborra A, Gierus M, Taube F, et al. Impact of grazing intensity on herbage intake, composition, and digestibility and on live weight gain of sheep on the inner Mongolian steppe. Livestock Science. 2009;**124**:142-147

[18] Serrano J, Shahidian S, Marques da Silva J. Calibration of GrassMaster II to estimate green and dry matter yield in Mediterranean pastures: Effect of pasture moisture content. Crop & Pasture, Science. 2016;**67**:780-791

[19] Cauduro GF, Carvalho P, Barbosa C, Lunardi R, Pilau A, Freitas F. A comparison of methods used to determine forage mass in Italian ryegrass pasture (*Lolium multiflorum* Lam.). Ciência Rural. 2006;**36**:1617-1623

[20] Serrano J, Peça J, Marques da Silva J, Shahidian S. Calibration of a capacitance probe for measurement and mapping of dry matter yield in Mediterranean pastures. Precision Agriculture. 2011;**12**:860-875

[21] Serrano J, Shahidian S, Marques da Silva J. Long-term evaluation of the Grassmaster II probe used to estimate productivity of dryland pastures. In: Stafford J, editor. Proceedings of the 14th European Conference on Precision Agriculture (Precision Agriculture '23, ECPA 2023); 2-6 July 2023; Bologna, Italy. Wageningen Academic Publishers. pp. 197-204

[22] Lugassi R, Zaady E, Goldshleger N, Shoshany M, Chudnovsky A. Spatial and temporal monitoring of pasture ecological quality: Sentinel-2-based estimation of crude protein and neutral detergent fiber contents. Remote Sensing. 2019;**11**:799

[23] Serrano J, Mendes S, Shahidian S, Marques da Silva J. Pasture quality monitoring based on proximal and remote optical sensors: A case study in the montado Mediterranean ecosystem. AgriEngineering. 2023;**5**:380-394

[24] Serrano J, Shahidian S, Paixão L, Marques da Silva J, Morais T, Teixeira R, et al. Spatiotemporal patterns of pasture quality based on NDVI time-series in Mediterranean Montado ecosystem. Remote Sensing. 2021;**13**:3820

[25] Serrano J, Shahidian S, da Silva JM. Monitoring seasonal pasture quality degradation in the Mediterranean montado ecosystem: Proximal versus remote sensing. Water. 2018;**10**:1422

[26] Pullanagari R, Yule I, Tuohy M, Hedley M, Dynes R, King W. Proximal sensing of the seasonal variability of pasture nutritive value using multispectral radiometry. Grass and Forage Science. 2013;**68**:110-119

[27] Sishodia RP, Ray RL, Singh SK. Applications of remote sensing in precision agriculture. A review. Remote Sensing. 2020;**12**:3136

[28] Huang S, Tang L, Hupy JP, Wang Y, Shao G. A commentary review on the use of normalized difference vegetation index (NDVI) in the era of popular remote sensing. Journal of Forestry Research. 2021;**32**:1-6

[29] Pezzopane JRM, Bernardi ACC, Bosi C, Crippa PH, Santos PM, Nardachione EC. Assessment of Piatã palisadegrass forage mass in integrated livestock production systems using a proximal canopy reflectance sensor. European Journal of Agronomy. 2019;**103**:1301-1339

[30] Handcock RN, Gobbett DL, González LA, Bishop-Hurley GJ, McGavin SL. A pilot project combining multispectral proximal sensors and digital cameras for monitoring tropical pastures. Biogeosciences. 2016;**13**:4673-4695

[31] Cicore P, Serrano J, Shahidian S, Sousa A, Costa JL, Marques da Silva J. Assessment of the spatial variability in tall wheatgrass forage using LANDSAT 8 satellite imagery to delineate potential management zones. Environmental Monitoring and Assessment. 2016;**188**:1-11

[32] Louhaichi M, Hassan S, Clifton K, Johnson DE. A reliable and non-destructive method for estimating forage shrub cover and biomass in arid environments using digital vegetation charting technique. Agroforestry Systems. 2018;**92**:1341-1352

[33] Serrano J, Roma L, Shahidian S, Belo A, Carreira E, Paniagua LL, et al. A technological approach to support extensive livestock management in the portuguese Montado ecosystem. Agronomy. 2022;**12**:1212

[34] Schaefer MT, Lamb DW. A combination of plant NDVI and LiDAR measurements improve the estimation of pasture biomass in tall fescue (Festuca arundinacea var. Fletcher). Remote Sensing. 2016;**8**:1-10

Chapter 7

# Sustainable Farming through Precision Agriculture: Enhancing Nitrogen Use and Weed Management

*Mehmet Hadi Suzer, Mehmet Şenbayram and Mehmet Ali Çullu*

## Abstract

The integration of digital tools to agriculture became more important than ever because of food security concerns and climate change. Real-time soil and crop monitoring systems, such as field sensors, spectral cameras, decision-making platforms and autonomous robots have significant potential to determine anomalies and optimize crop management practices. For example, variable rate application methods consist of reliable vegetation cover maps, however, do not contain information about the underlying causes of variation. Thus, the benefits of precision management remain a subject of debate limiting the adoption of such technology by farmers. In this review, we discuss the underlying causes of lower success rates of variable rate application and the developing of new digital platforms which will improve the efficiency of digital farming tools to manage nitrogen. Furthermore, image-based weed detection (key milestone for digitalized weed management) that employs sophisticated algorithms and machine learning techniques to analyze images captured by drones or ground-based cameras to identify weed species, density, and its growth stages, enabling targeted weed control will be discussed. Adoption of upcoming digital tools not only contributes to a significant technological leap in agriculture, but we believe also be the most important drivers of sustainable agriculture.

**Keywords:** digital farming, weed detection, nitrogen, NDVI, satellite

## 1. Introduction

Agriculture plays a vital role in meeting the global demand for food and fiber, and in achieving socio-economic balance [1]. As the world's population grows, there is an increasing demand for higher rates of food production on a global scale. However, with limited arable land and natural resources, increasing food production to feed the estimated 9 billion people by 2050 will be the biggest challenge for the human race [2]. The application of nitrogen fertilizers is fundamental for high crop production, but only a portion is utilized by plants [3, 4]. The excessive use of nitrogen-based fertilizers worldwide has caused a surge in nitrate pollution of water sources [5, 6], and increased atmospheric nitrous oxide emissions [7, 8].



IntechOpen

Excessive use of chemical fertilizers leads to their accumulation in soil, which negatively impacts crop quality and yield, and causes environmental pollution [9, 10]. Around 175.5 million tons of chemical fertilizers were utilized in farming to produce optimal crop yield [11]. Approximately 60% of the total anthropogenic $N_2O$ emissions come from agricultural soils [12–14], making it a significant contributor to climate change. Nitrous oxide ($N_2O$), which has 265 times more global warming potential than $CO_2$, contributes to greenhouse gas (GHG) emissions [15]. The Food and Agriculture Organization [7] predicts a 50% increase in worldwide usage of synthetic N fertilizers by 2050 compared to 2012. Reducing $N_2O$ emissions per unit of nitrogen compound applied can be a potential strategy to minimize the global impact of nitrogen fertilizers on human-induced GHG emissions. However, the most effective approach is to optimize nitrogen fertilizer management, where excess fertilization is one of the biggest problems of sustainable agriculture globally [16, 17].

Integration of advanced sensors and control systems together with advancements in artificial intelligence have significantly accelerated the evolution of agricultural robotics, where we already see practical implementations of these technologies in advanced farmers' fields. Change from traditional agriculture to fully automated farm models allows for acquiring accurate and detailed temporal and spatial information on time that can be used to complete nonlinear control tasks for almost every agricultural management practice by autonomous vehicles [18]. The potential of fully- or semi-autonomous robot systems applications attracts not only farmers but also global stakeholders, engineers from various classes, and scientists. Although most of the fully autonomous field robots are in prototype phase, some of them are already in the market and now capable of performing various tasks such as crop inspection [19], sowing and weed control [20], harvesting [21], targeted spraying [22], optimized nitrogenous fertilizer management [23].

The surge in big data's role is reshaping agricultural technology. Farmers now have access to a wealth of information, from tractor display screens to mobile farm management applications. This information not only aids in decision-making but also feeds into the development of new agricultural technologies by companies. These firms leverage the collected data to refine digital farming solutions and innovate in areas such as fertilizer and agrochemical management. The landscape of digital farming innovation is heavily influenced by start-up firms, which are key players amidst a global food economy. Large corporations are actively seeking to lead in the digital farming arena, either by developing their technologies or acquiring smaller companies. However, this rapid advancement and integration of data collection and analytics in agriculture have resulted in fragmented and ambiguous governance structures. The regulation of data collection and access in digital farming is predominantly governed by user agreements formulated by the developers, which farmers are required to sign. In reaction to these developments, there is a growing movement among farmers advocating for data ownership rights.

Sensors in the agricultural field monitor environmental factors such as humidity, temperature, and wind, while images (either photos or videos) taken from soil and crops can provide information about crop development, diseases, weeds, and harmful insects [24–28]. Since the healthy development of the plant determines the benefits to be obtained and the amount of resources to be spent, traditional human-eye monitoring is subject to timeliness and accuracy issues. Real-time and accurate monitoring of the plant is a crucial component of precision agriculture, playing a significant role in maximizing crop yield [14].

## 2. Fertilizer management

### 2.1 Remote sensing nitrogen

The integration of digital farming tools in fertilizer management within croplands has transformed agricultural practices, particularly in terms of precision and sustainability. Digital farming, a concept dating back nearly a century, has evolved significantly with technological advancements. The modern form, driven by data analytics, emerged in the 1980s and 1990s, incorporating tools like GPS, yield monitor mapping, and variable-rate application of agrochemicals [29]. Recent developments include machine learning (ML) tools, remote sensing, advanced variable-rate technologies, robotics, auto-steer machinery, and unmanned aerial vehicles. These technologies enable site-specific and global data collection and analysis, influencing decisions on most agricultural practices (e.g., fertilizer management), contrasting the uniform practices dominant since the introduction of agrochemicals a century ago. The adoption of these technologies is not only prevalent on large commodity crop farms in industrialized countries but now available also for small farmers all over the globe. The deployment of these technologies is facilitated by the decreasing cost of digital technology and advances in computing and robotics, enhancing their accessibility and affordability.

*Simple apps showing time course of satellite-based NDVI data for farmers:* The time course of satellite-based normalized difference vegetation index (NDVI), when integrated into a simple computer software or a phone app, can significantly impact fertilizer management for all farm types in several profound ways. NDVI values can be used as a graphical indicator that assesses whether the target being observed contains healthy green vegetation. By using satellite-based NDVI imagery, farmers can obtain detailed insights into the health and vigor of their crops. This data may reflect the amount of chlorophyll present, which can be linked to the growth stage of their crops (for timing of fertilization) or in some way to the nitrogen uptake of the plant canopy for estimating the N input needed. By analyzing these images, farmers can estimate the growth stages of their crops more accurately and assess whether the plants are getting enough nitrogen or if they are experiencing deficiencies in any part of their field. Furthermore, anomalies in plant growth and nitrogen uptake can be effectively detected, allowing for rapid actions. This approach to crop management can lead to a more efficient use of resources, cost savings, and a reduced environmental $CO_2$ footprint, allowing significant advancement in sustainable agricultural practices.

**Figure 1** represents NDVI values for wheat fields with different sowing dates, as captured by a satellite and processed through an NDVI satellite app (https://maps.datafarming.com), on January 7, 2024. NDVI is a measure of the health and vigor of vegetation, with values ranging from −1 to +1, the higher the value, the healthier the vegetation. We can see three distinct sections in **Figure 1**, each corresponding to a different sowing date and displaying a unique color range that indicates the NDVI value hence indicating different growth stages of the wheat. The top section shows a field sown on November 15, 2023 with an NDVI value corresponding to the BBCH scale score of 22, which generally indicates tillering in wheat. The field section sown on December 4 corresponded to BBCH of 13 and the field sown on December 17 corresponded to BBCH of 9, suggesting that the wheat is at the emergence. The color gradient from blue (high NDVI values) to red (low NDVI values) visually demonstrates the difference in vegetation density and health across the fields, with the fields

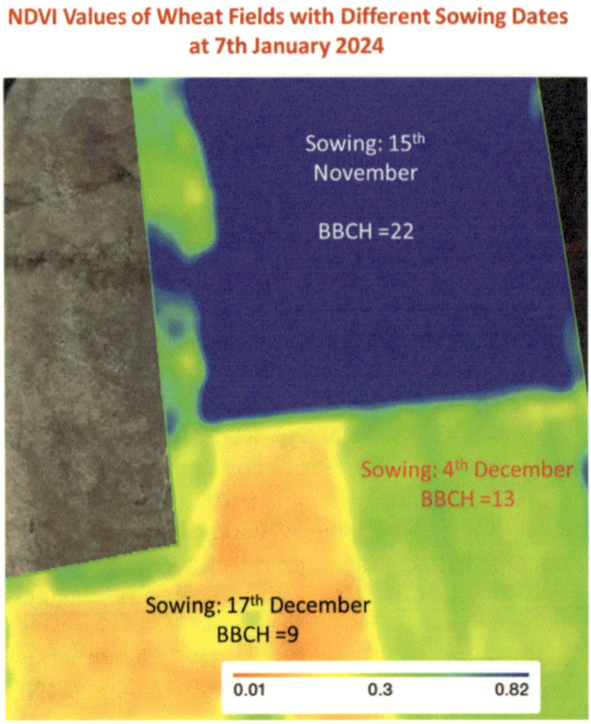


**Figure 1.**
*NDVI values, acquired from sentinel satellite imagery, reveal various developmental stages of wheat crops. This data offers farmers a comprehensive view of their wheat fields' growth progress.*

sown earlier showing higher NDVI values indicating more advanced growth stages. This app provides farmers with reliable information on the growth stages of their wheat, which is crucial for making precise decisions about, for example, to the timing of the first fertilizer application.

Nonetheless, the availability of satellite imagery can be inconsistent, as conditions often obstruct the collection of clear data specifically in winter or even during summer in northern Europe. Cloud cover can render satellites ineffective on certain days, leading to gaps in monitoring and potential delays in agricultural decision-making. Drones equipped with spectral cameras represent a significant innovation in precision agriculture, offering high-resolution imaging capabilities that surpass those of many satellite systems. These cameras capture data across multiple spectral bands, not just the visible spectrum, enabling detailed analysis of crop health and soil conditions through indicators such as NDVI. The high-resolution imagery from drones can reveal intricate details of field conditions, such as plant counting, pest infestation levels, and even nutrient deficiencies at a very granular level. For some farmers, especially those managing large-scale or highly specialized operations, this level of detail can be invaluable, allowing for highly precise interventions that can improve yields and reduce input costs. However, not all farmers require this high degree of resolution. For many agricultural applications, the detail provided by satellite imagery may be sufficient, especially when the focus is on broader field trends rather than minute variations. The added cost and complexity of drone operations, including the need to process large volumes of data, may not justify the incremental benefits for these users. Moreover, drones offer a distinct advantage over satellite imagery in terms of flexibility and reliability under different weather conditions. Satellite imagery is dependent on the satellite's orbit and is subject to atmospheric conditions, meaning that cloud cover can significantly impede the acquisition of useful data. Drones, on the other hand, operate at lower altitudes and can often fly below cloud level, providing more consistent data collection opportunities, especially in regions with frequent cloud cover or during seasons where sunny days are scarce. In summary, while drones with spectral cameras offer higher-resolution data, the necessity and practicality of this technology for individual farmers depend on the scale, precision requirements, and specific challenges of their agricultural operations. The ability to operate in less-than-ideal weather conditions is a clear advantage of drones, ensuring that critical agricultural decisions are informed by the most current and precise data available, regardless of cloud cover.

Soil sensors are vital tools in precision agriculture, enabling farmers, for example, to monitor soil water, temperature and even pH changes precisely. Such real-time data can lead to more precise water and nutrient management, ensuring that crops receive the right amount of water and nutrients. However, there is a notable gap in the market for reliable remote sensors specifically for nitrogen in soil. While there are sensors that can approximate nitrogen availability through indirect measures, the complexity of soil chemistry and the multitude of factors influencing nitrogen availability make it challenging to accurately measure soil nitrogen levels remotely. This difficulty is compounded by the fact that nitrogen is highly mobile within the soil, existing in various forms and undergoing numerous transformations.

In contrast, handheld chlorophyll meters such as SPAD (soil plant analysis development) meters, provide a non-destructive and reliable means of assessing leaf chlorophyll, which correlates closely with nitrogen content in plant tissue. By measuring the greenness of the leaf, SPAD meters give a quick and reliable indication of the plant's nitrogen status. This method is widely accepted in agronomic practices as a proxy for leaf nitrogen content, and many studies have validated its use. While remote sensing technologies are advancing, the precision and reliability of direct measurements like SPAD for nitrogen management remain unmatched. Until remote sensors can measure soil nitrogen directly and reliably, tools like SPAD meters will continue to be essential for farmers aiming to optimize their nitrogen fertilizer use. It is important for ongoing research to focus on developing remote sensing technologies that can accurately reflect soil nitrogen levels, bridging the current gap and enabling even more efficient and environmentally friendly farming practices.

## 2.2 Robots in agriculture

*Crop health monitoring:* Accurately mapping the spatial distribution of crop development throughout the whole vegetation period is crucial for monitoring plant stress and crop health. Satellites, human-crewed aircraft, drones or ground-based autonomous vehicles equipped with spectral, thermal or RGB cameras are popular remote-sensing technologies [30]. Proximal canopy sensors measure reflected light at specific wavelengths and convert these to various vegetation indexes (e.g., NDVI may, for example, represent the soil coverage of photosynthetically active biomass [31, 32]. These indices can be used to estimate crop biomass, N nutritional status or yield potential, and several commercial sensors are available as well as different methods to transform their readings into N rates. The absorption of light in the visible region of the spectrum (450–680 nm) by photosynthetic pigments such as chlorophyll a and b, or carotenoids, has been well documented [33], and the content of these pigments are known to be affected by the crop health. The "red-edge" region (680–780 nm), lying between the visible and near-infrared (NIR) regions, is known to be related to the levels of chlorophyll content which can be used as an indicator of crop productivity [34]. The NDVI is a widely used spectral index that measures the difference between the reflectance values of near-infrared and red wavelengths and is positively correlated with crop yield and health. Other spectral indices, such as the green normalized difference vegetation index (GNDVI), soil-adjusted vegetation index (SAVI), and plant senescence reflectance index (PSRI), have also been used to screen vegetation health and may additional information about the soil/vegetation ratio [35]. Although RGB pictures can also be used for crop health monitoring (discussed in Section 3 in more detail), spectral cameras are simpler solutions as they do not require sophisticated ML algorithms for simple monitoring processes. Here we listed the advantages and shortcomings of various mobile remote sensing platforms (**Table 1**).

*Fertilizer application:* Ground vehicles equipped with crop sensors, GPS, and advanced mapping systems can be used for high-precision variable rate application of nutrients such as nitrogen [32]. This approach may optimize the nitrogen needs of specific areas within a field that show varying growth patterns. However, the demand and the supply of N is a complex task as the optimum N rate depends on numerous interrelated factors such as the expected yield, and the N supply from other sources besides fertilizer and N losses to the environment [36]. Variable rate maps, segmented according to different crop development levels, can be created at high resolution through (i) offline methods such as satellite imagery (utilizing either current or long-term NDVI data analysis), or unmanned aerial vehicles (e.g., DJI Mavic 3 Multispectral, Parrot Bluegrass), or (ii) online methods include ground vehicle-attached multispectral sensors (e.g., tractor-mounted sensors like the YARA N-sensor and CropSpec by Topcon). Although these tools can generate reliable maps of the field segmented by different growth rates, background information for the cause of such variation is still missing to define whether a higher or lower rate of fertilizer is needed for each segment. This is because the heterogeneity in growth rate at specific locations in the field may not always directly relate to nitrogen limitations in those specific spots [37]. Therefore, the benefits of such precision management practices remain still a subject of debate specifically when it comes to crop yield. For example, in a review study, authors concluded that N fertilizer savings of 5–45% with no significant effect on yield, and lack of consistent

| | **Advantages** | **Shortcomings** |
|---|---|---|
| Satellite | Wide coverage (field to regional screening), low cost, no pilot involvement, long-term data analysis | On average 4–10 days data availability, weather dependent (no data in cloudy days), limited resolution |
| Drones | High resolution, easy deployment and setup | Limited flight times, pilot involvement, average payload weight, limited real-time data analysis, limited battery life |
| Fixed wing aircrafts | Large area coverage (faster than drones), high resolution | Pilot involvement, higher risk for collusions, complicated set-up, less maneuverable than drones, lower resolutions than drones |
| Ground-based vehicles | Very high resolution, allowing for real-time data analysis, lower risk for accidents and collusion | Limited to accessible areas, slower data collection, higher costs, sophisticated algorithms for row crops |

**Table 1.**
*Advantages and shortcomings of various mobile remote sensing platforms.*

evidence of economic benefits limits the adoption of variable rate technology by farmers [38]. Similarly, several other authors concluded no clear evidence that using sensors was profitable, because in most of their field observations reduced N rates often negatively affected yield, leading to smaller profits [39–41]. These studies indicate that additional data (such as soil texture, weather data, availability of other nutrients, and historical crop yields) is crucial for a more comprehensive understanding of nitrogen needs in varying field segments. For example, integrating historical yield, or potential yield maps generated from multi-year satellite images, can help in understanding the variation in soil fertility of different field segments. This variation may indicate chronic problems such as soil texture, biology, or physical orientation (which affect solar radiation absorption) that lead to growth depression in specific locations, highlighting the need for further analysis. Here, soil analysis can reveal a nutrient profile and identify deficiencies in nutrients other than nitrogen, or texture and organic matter issues for each varying field segment. Incorporating these complex datasets requires sophisticated data analysis seeking inter-relations and expertise, pointing to a collaboration between agronomists, data scientists, and farmers. Once complex data sets are achieved, advanced data analytics and ML algorithms can then be employed to develop more accurate algorithms for optimizing the timing and rate of nitrogen fertilizer applications. However, as far as we are aware, we are still in the initial phase of the development of effective digital farming. However, with the development in sensor and automation technologies in conjunction with the sophisticate ML processes for big data analysis, we expect that precision agriculture will likely become more and more accurate and effective. In conclusion, while current tools provide a significant step forward in precision agriculture, a holistic approach that incorporates a wider range of data sources and advanced analytical methods is essential. The future of precision agriculture lies in the seamless integration of various data streams, enabling farmers to make more informed decisions that lead to sustainable and efficient agricultural practices.

*Greenhouse gas measurements:* Considering the major source of GHG emissions is agriculture, mobile GHG measurement systems is very important to identify the spatial variation and the annual quantity of GHG emission from agricultural soils. Systems that can measure GHG emissions such as ($N_2O$, $CO_2$, and $CH_4$) in the concentration range relevant for cropped soils are already available. For instance, the Picarro cavity ringdown spectroscopy can measure $N_2O$, $CO_2$, and $CH_4$ at very high precision, however, this system is very expensive and with very high power consumption. Other systems, such as GASERA's ONE GHG, which uses photoacoustics have the necessary accuracy or most GHGs ($N_2O$, $CO_2$, and $CH_4$) in the ppb range, field robots can be fitted with such sensors to detect and measure GHG emissions directly from the soil and plants with very high time resolution and accuracy without the need for excess manual work. These measurements are crucial for understanding the interactive effect of environment and agricultural practices for developing site-specific strategies to mitigate GHG emissions.

In this context, the robot shown in **Figure 2** has been developed by the scientific team at the University of Harran to determine GHG emissions in agricultural soils to determine the impact of various management practices such as N fertilization, irrigation and tillage. The robot has various optical sensors (Intel Realsense depth camera) and GASERA ONE, Finland, developed for real-time monitoring of GHG emissions and crop development. This robot can accurately measure the concentrations of emitted gases like methane ($CH_4$), carbon dioxide ($CO_2$), ammonia ($NH_3$), and nitrous oxide ($N_2O$) at very high time resolution, which are relevant for determining environmental impact of farming practices. We can conclude that the future of digitally

**Figure 2.**
*Fully autonomous field robot measuring GHG emission from crop fields developed by the team at the University of Harran.*

optimized nitrogen fertilizer management does not rely only on determining crop N needs. It necessitates the integration of sophisticated data analysis, encompassing factors like gaseous nitrogen losses, weather forecasts, soil parameters, and real-time crop monitoring. This analysis should ideally be performed by advanced ML techniques to develop reliable algorithms for determining timing and rate of fertilizers needed for sustainable crop production. However, significant progress and development in this field are still in its early steps.

## 3. Machine learning algorithms for precision agriculture applications

Monitoring the vast agricultural lands plays a crucial role in obtaining continuous, real-time, and accurate information about the soil and plants, directly impacting the quality and quantity of the harvested crop. Simultaneously, optimizing resource usage helps reduce both costs and harmful environmental effects. However, due to reasons such as the low precision and accuracy of information gathered through human observations and the high labor costs of, especially very large fields, computer vision systems have become a necessity. Computer vision systems can acquire information about the surrounding physical environment through their sensory and imaging devices, interpret this information, and transform it into useful outputs.

Modern digital detection systems can obtain high-resolution and real-time data from a variety of sensors and imaging devices and utilize advanced analysis methods for processing this information. Analyses that were traditionally performed using image processing techniques have recently been transitioning to more sophisticated and advanced ML algorithms. A subset of ML algorithms known as deep learning (DL), can efficiently address various image separation and classification problems. Therefore, these algorithms hold great promise for the analysis, interpretation, and generation of useful outputs from large agricultural datasets of today.

Utilization of ML algorithms in the analysis of images taken from plants or soil in agricultural fields has resulted in significant improvements in object detection, identification, and classification [24]. ML methods are applied not only to standard red, green, and blue (RGB) images but also to images captured in different color spectra such as near infrared (NIR) and color infrared (CIR). ML algorithms generally involve a learning process centered around "experience," where a model is trained with "training data" and later used for classifying and extracting features from the test data. ML tasks can be classified based on the learning method, as supervised and unsupervised, and the learning model, as classification, regression, clustering, and dimensionality reduction.

In supervised learning, labeled input and output examples are provided to create a model representing the relationship between the input and the output. The trained model is then used to infer useful information from the test data. In unsupervised learning, data is explored for hidden features (patterns) without human verification, as the data is not labeled as training or testing. Reinforcement learning uses a reward mechanism instead of labels in supervised learning, where each verified output is rewarded, and the goal is to maximize the reward amount. Reinforcement learning does not require large datasets.

### 3.1 Data preprocessing for deep learning

In the application, images obtained with cameras or scanners undergo preprocessing such as filtering and segmentation after which various texture and color features are obtained. These features are processed by an artificial neural network (ANN) classifier to classify the image. We give a summary of data preprocessing stages, which significantly improve data for further analysis. Preprocessing of data improves accuracy while reducing computational resource demand of ML algorithms.

#### *3.1.1 Data acquisition*

ML algorithms require a training process before being tested on real-world problems. They need a sufficient amount of example data to train themselves with the phenomenon they will be testing. Example data can be raw data collected by digital sensors, or visual data obtained from cameras on satellites, drones, field robots (such as the one shown in **Figure 2**), or portable cameras. For further analysis, these images can also be examined in different spectrums such as RGB, NIR, or CIR. Additionally, data go through preprocessing via classical digital signal processing techniques (such as fast Fourier transform, FFT, finite impulse response, FIR, etc.) or image processing techniques (edge detection, thresholding, etc.) to enhance the efficiency of ML algorithms.

#### *3.1.2 Data labeling*

For the training of ML models, visual data obtained from various sources are labeled to known descriptors. In this stage, objects in regions (Region of Interest, ROI) distinguishable in the data are labeled, allowing the ML algorithm to build a model for each object. During the test phase, the ML algorithm is run on unlabeled data, and objects are identified and classified using the trained model. Labeling can be done manually or automatically, and the presence of an existing database of labeled data may eliminate the need to repeat this stage.

#### *3.1.3 Data augmentation*

For the training of ML models, a large quantity and number of real datasets are required. However, often, it may not be feasible to create such a dataset due to constraints such as time, labor, and cost. In such situations, augmenting the existing dataset through synthetic means can mitigate this problem. To achieve this, the dataset can be enriched with various transformations such as zooming, cropping, flipping, rotating, overlapping, and color changing, etc.

### 3.2 Major ML algorithms for object detection

ANNs draw inspiration from the functions of the human brain and mimic complex functions such as object recognition, learning, and decision-making. Similar to the billions of interconnected and communicating neurons in the human brain, an ANN consists of numerous processing units that are organized in a certain way and connected to each other.

As AI algorithms work on massive amounts of data, they are resource hungry in terms of computational power, memory, communication bandwidth, etc. There have been continuing efforts to increase their accuracy and optimize their resource usage further. Recently, another ML family of algorithms, the deep neural networks (DNN), has got a lot of attention for their successful implementations in object recognition and classification. DNN algorithms consist of multiple hidden layers, and convolutional and pooling layers as well, compared to the single hidden layer in ANNs.

In DNN algorithms, the feature extraction stage is performed by the model itself. The most popular deep learning algorithms capable of mapping features on an image are convolutional neural networks (CNN) and its derivatives (RCNN, fast RCNN, faster RCNN), long-short term memory (LSTM), and single stage algorithms, YOLO, and SSD.

These DNN algorithms are described briefly as follows.

#### *3.2.1 Convolutional neural networks (CNN)*

The CNN model is specifically developed for image processing and recognition tasks. It supports transfer learning, the capability of retraining the model over the existing data for new recognition tasks. Additionally, various derivatives of the CNN model are developed, which require less image preprocessing due to their ability to perform feature mapping (**Figure 3**).

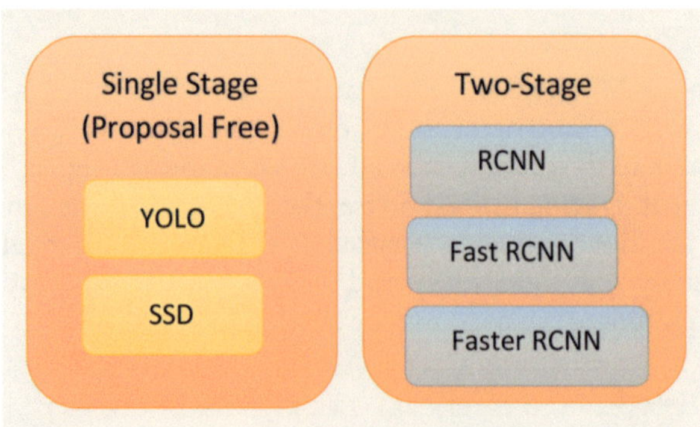


**Figure 3.**
*ML algorithms for object detection.*

Instead of the single type of hidden layer in ANN architecture, CNN consists of three types of layers. The first one is the convolution layer, in which there are filters acting as neurons. Feature maps are obtained at the output of this layer by convolving the filters with the source image. These feature maps perform tasks such as edge detection and object boundary detection which would otherwise need to be obtained at the preprocessing phase.

Pooling layers, on the other hand, aim to reduce the computational resource requirements of the algorithm by generalizing the obtained feature maps and reducing their dimensions without incurring any loss.

Finally, in the "fully connected layer," feature maps from the pooling layer are flattened and the classification of the image is performed using these flattened feature maps.

#### *3.2.2 Long-short term memory (LSTM)*

LSTM networks are a type of recurrent neural network (RNN) designed to detect long-term dependencies in sequential data [42]. LSTM networks have a memory cell that can maintain information over long periods of time. They can generate variable-length outputs from variable-length inputs. Many variants such as CNN-LSTM and stacked LSTM have been developed enhancing the performance and capabilities further.

#### *3.2.3 Deep recurrent Q-network (DQN)*

DQN networks combine the Q-learning algorithm (which aims to maximize long-term benefits) and DNN networks, with the aim to predict the expected benefit for a specific behavior. DQN networks have been successfully applied in applications such as robotic control, autonomous vehicles and have also been used for crop yield prediction in agriculture [43].

#### *3.2.4 Region-based CNN (R-CNN)*

It is a two-stage CNN method developed specifically for object detection [44]. As illustrated in **Figure 4**, in the first stage, the image is segmented into regions of interest (ROI) based on features such as texture and color. Then, each ROI is fed to a CNN

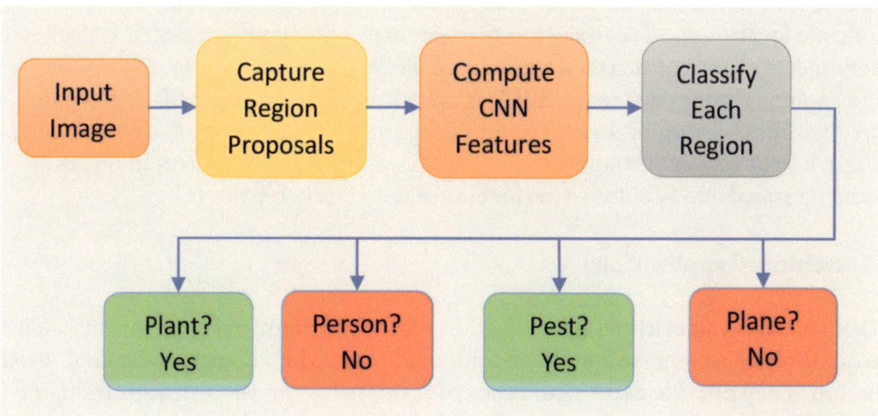


**Figure 4.**
*R-CNN algorithm overview.*

layer to create a fixed-length feature vector. CNNs consist of pre-trained networks. Although R-CNN has achieved quite successful results (in terms of accuracy) in its time, it has remained relatively slow in both the training and testing phases.

#### *3.2.5 Fast R-CNN*

Developed to accelerate the R-CNN algorithm, fast R-CNN creates a single feature map for all ROIs that can be shared among all, instead of having a separate feature extraction step for each ROI [45]. It provided more accurate and much faster results compared to R-CNN.

#### *3.2.6 Faster R-CNN*

Faster R-CNN is even more improved version of the fast R-CNN algorithm in terms of speed and accuracy [46]. For each position on the image, a *region proposal network* (RPN) predicts objectness scores and object bounds at the same time. RPNs are trained to generate almost cost-free region proposals which are to be fed into fast-RCNN for object detection. Together with fast-RCNN, RPNs achieve 5fps speed and 70.4% mAP (mean average precision) object detection accuracy.

#### *3.2.7 YOLO*

You only look once (YOLO) is an object detection method that analyzes images in a single pass, providing both object classification and object location. YOLO utilizes an end-to-end neural network that makes predictions of bounding boxes and class probabilities in one evaluation. This allows it to achieve 45 frames per second (fps) image processing [47] which makes it effective in fast, real-time applications. YOLO continues to improve through a series of versions up to YOLO v4 (as of this date) and each new version improves performance further, up to 72 fps and 88% mAP for YOLO v4 [48].

#### *3.2.8 SSD*

Single shot multi-box detector (SSD) is a model that can perform object detection at multiple locations and scales using feature maps of different scales. It is designed to detect objects of different sizes simultaneously. It achieves 59 fps speed and claimed to be simpler as it does not require object proposals which makes SSD to be easily trained while achieving higher accuracy compared to other single-stage methods [49].

**Figure 5** shows performance of YOLO, SSD, and CNN derivatives in terms of processing speed and accuracy (performance data is taken from [48–50]).

### 3.3 Agricultural applications

Deep learning algorithms are utilized in various agricultural applications such as crop quality assessment, soil nutrient and yield prediction, disease, pest, and weed detection. They provide expert guidance to farmers for the amount and timing of additional nutrients, irrigation, chemicals for disease control, pesticide, and herbicides which increase yield quality and amount.

We will briefly describe these application areas as follows.

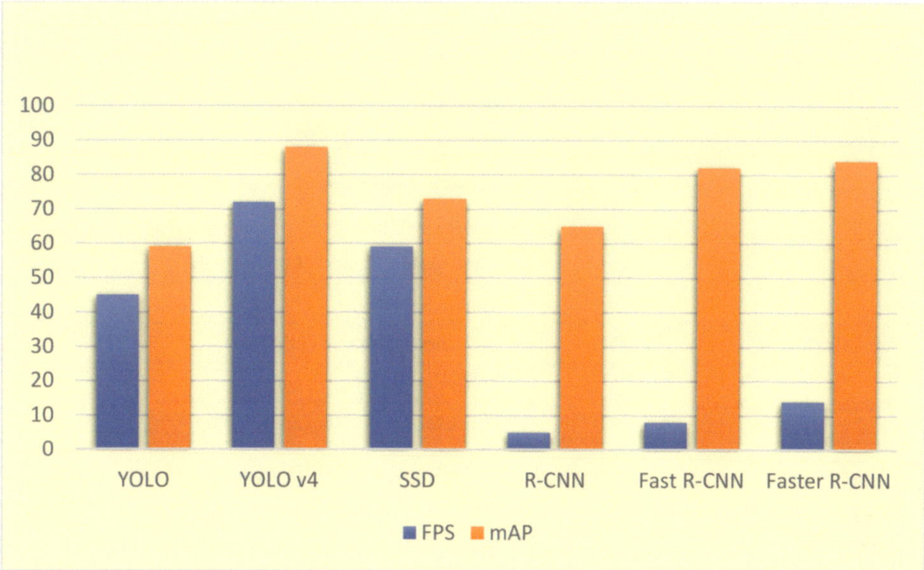


**Figure 5.**
*Performance comparison of DNN algorithms for object identification.*

### *3.3.1 Crop quality assessment and yield prediction*

Assessing crop quality involves determining the number, weight, size, and maturity level of crops during plant growth. Assessment is performed with ML methods over raw images, thus requiring no direct contact (non-invasive), eliminates manual assessment, and reducing labor while achieving higher precision. Accurate assessment of crop quality using computer vision and ML together can also enable automated harvesting.

Yield prediction in early stages of plants' growth is another area where ML techniques play a significant role. Accurate yield prediction can contribute to food production and food security by optimizing resources such as water and nutrients etc. Yield prediction accuracy depends on many factors involved in plant life cycle and requires managing very large datasets with ML algorithms.

Studies using support vector machine (SVM), Bayesian model, Gaussian Naive Bayes (BM/GNB), artificial neural network, adaptive neuro-fuzzy inference system (ANN/ANFIS), artificial neural network, self-organizing neural networks (ANN/SNKs), and clustering, expectation-maximization (EM) models have been conducted, incorporating drone, satellite, and camera images. **Table 2** shows several features used to predict the crop yield [51].

### *3.3.2 Disease/pest detection*

Plant diseases and organisms such as harmful insects and parasites (pests) living on plants negatively affect crop production. Detecting the symptoms of diseases and pests with the naked eye is quite challenging (see **Figure 6**) and has low accuracy particularly during the early stages of plants' growth. The delayed identification of diseases and pests leads to the need for excessive drug/pesticide usage in later stages, reducing crop quality and increasing environmental harm. Therefore, the timely and accurate detection of diseases and pests is of critical importance to determine the type and dosage of drugs or pesticides to be applied.

| | |
|---|---|
| *Soil information* | Soil type |
| | Soil pH level |
| *Field information* | Irrigation |
| | Fertilization |
| *Weather information* | Ambient temperature |
| | Rainfall and precipitation |
| | Wind speed and atmospheric pressure |
| | Solar exposure time |
| *Nutrient information* | Nutrients available and injected |
| *Artificial information* | Enhanced vegetation index (EVI) |
| | Normalized vegetation index (NDVI) |

**Table 2.**
*Features used for crop yield prediction [51].*

**Figure 6.**
*Leaves with disease and pests.*

Traditionally, pesticides are uniformly sprayed on cultivated areas for pest and disease control. Although effective, this method leads to high costs and environmental damage, including the mixing of pesticide residues with crops and drinking water, and negative effects on wildlife and ecosystems. Unlike traditional methods, ML methods ensure that the necessary pesticide is applied only at the required time and amount to the infected plant, minimizing unnecessary financial and environmental impacts. In the literature [27, 52], SVM, ANN, CNN, YOLOv3, LSTM and their derivatives (explained in Section 3.1) have been used for pest and disease identification and classification.

**Figure 7.**
*Weed detection over RGB image.*

### *3.3.3 Weed detection*

Weeds (**Figure 7**) pose a significant threat to crop production, as they share resources such as soil, water, nutrients, sunlight, etc., reducing crop quality. Detecting and distinguishing weeds for targeted herbicide applications is challenging. Contrary to traditional methods, such as manual or chemical weed control methods, ML algorithms allow real-time detection, identification and classification of weeds and their effective destruction, minimizing the side effects of herbicides and their negative impact on the environment. Algorithms such as ANN, CNN, and SVM derivatives are well suited to weed detection applications.

## **4. Conclusion**

The integration of NDVI satellite data and drone technology, complemented by ground sensors (e.g., SPAD), has established a new standard in nitrogen fertilizer management. These advanced tools enable a dynamic and responsive approach to fertilization that aligns closely with the crop nutrient status. Satellite-derived NDVI provides a macroscopic view of field variability and plant health, allowing for the identification of areas that may require differential nitrogen application. Drones fill in the gaps on cloudy days and add a layer of detail with their high-resolution imagery, capable of monitoring crop health at a micro-scale, even detecting individual leaves' nutrient status. Handheld devices like SPAD meters complement these technologies by offering on-the-spot assessments of leaf chlorophyll, serving as another dependable indicator of plant nitrogen content. The confluence of NDVI data, drone imaging, and SPAD readings supports farmers in applying nitrogen fertilizers more effectively, matching the spatial and temporal crop needs. However, as far as we are aware, we are still in the initial phase of the development of effective digital farming.

With the development of sensor and automation technologies in conjunction with the sophisticated ML processes for big data analysis, we expect that precision agriculture will likely become more and more accurate and effective. The breakthroughs in image-based weed detection is one of the biggest breakthrough in weed management practices. Leveraging complex algorithms and machine learning, this technology facilitates discerning weed species and their respective growth stages, thereby enabling farmers to execute more focused and efficient weed control strategies. By synthesizing data from soil maps, yield maps, and remote sensing, these innovations ensure that each field zone receives just the right amount of input, fostering a reduction in herbicide usage, a decrease in production costs, and a lesser environmental impact from crop protection chemicals.

## Conflict of interest

The authors declare no conflict of interest.

## Author details

Mehmet Hadi Suzer[1], Mehmet Şenbayram[2] and Mehmet Ali Çullu[2*]

1 Department of Computer Engineering, Harran University, Şanlıurfa, Türkiye

2 Department of Soil Science and Plant Nutrition, Faculty of Agriculture, Harran University, Şanlıurfa, Türkiye

*Address all correspondence to: macullu@harran.edu.tr

## References

[1] UNFCCC. UNFCCC, 2015. United Nations: Paris Agreement; 2015. Available from: https://unfccc.int/sites/default/files/english_paris_agreement.pdf

[2] FAO. The future of food and agriculture - trends and challenges, food and agriculture Organization of the United Nations. Channels. 2017;**4**:180

[3] Reay D. Nitrogen and Climate Change: An Explosive Story. London: Palgrave Macmillan; 2015. DOI: 10.1057/9781137286963

[4] Menegat SAL, Ledo L, Tirado R. Greenhouse gas emissions from global production and use of nitrogen synthetic fertilizers in agriculture. Scientific Reports. 2022;**12**:14490. DOI: 10.1038/s41598-022-18773

[5] Singh B, Craswell E. Fertilizers and nitrate pollution of surface and ground water: An increasingly pervasive global problem. SN Applied Sciences. 2021;**3**:518. DOI: 10.1007/S42452-021-04521-8

[6] Chiu MT, Xu X, Wei Y, Huang Z, Schwing AG, Brunner R, et al. Agriculture-vision: A large aerial image database for agricultural pattern analysis. In: Proceedings of the IEEE/CVF Conference on Computer Vision and Pattern Recognition. Seattle, WA, USA: IEEE; 2020. pp. 2828-2838

[7] FAO. FAOSTAT. Emission shares dataset. 2022. Available from: fenix.fao.org/faostat/internal/en/#data/EM [Accessed: Mar. 11, 2022]

[8] Davidson EA, Keller M, Erickson HE, Verchot LV, Veldkamp E. Testing a conceptual model of soil emissions of nitrous and nitric oxides. Bioscience. 2000;**50**:667

[9] Gan L. Environmental risks of fertilizer use and the prevention and control measures in Chinese rural areas. Chimica Oggi-Chemistry Today. 2016;**34**(6 B):33-38

[10] Ajeng AA, Abdullah R, Malek MA, Chew KW, Ho Y-C, Ling TC, et al. The effects of biofertilizers on growth, soil fertility, and nutrients uptake of oil palm (Elaeis Guineensis) under greenhouse conditions. PRO. 2020;**8**:1681. DOI: 10.3390/pr8121681

[11] FAO. World Fertilizer Trends and Outlook to 2022. Rome, Italy: Food and Agriculture Organization of the United Nations (FAO); 2019

[12] Hu HW, Chen D, He JZ. Microbial regulation of terrestrial nitrous oxide formation: Understanding the biological pathways for prediction of emission rates. FEMS Microbiology Reviews. 2015a;**39**:729-749. DOI: 10.1093/femsre/fuv021

[13] Clark MA, Domingo NGG, Colgan K, Thakrar SK, Tilman D, Lynch J, et al. Global food system emissions could preclude achieving the 1.5° and 2°C climate change targets. Science. 2020;**370**:705-708. DOI: 10.1126/science.aba7357

[14] Tian H et al. A comprehensive quantification of global nitrous oxide sources and sinks. Nature. 2020;**586**:248-256. DOI: 10.1038/s41586-020-2780-0

[15] IPCC. Climate change and land: An IPCC special report on climate change, desertification, land degradation, sustainable land management, food security, and greenhouse gas fluxes in

terrestrial ecosystems, summary for policymakers. 2019

[16] Sutton MA et al. Our nutrient world: The challenge to produce more food and energy with less pollution. Global Overview of Nutrient Management. Available from: www.initrogen.org and www.gpa.unep.org/gpnm. 2013 [Accessed: Mar 11, 2022]

[17] Lassaletta L et al. Nitrogen use in the global food system: Past trends and future trajectories of agronomic performance, pollution, trade, and dietary demand. Environmental Research Letters. 2016;**11**:095007. DOI: 10.1088/1748-9326/11/9/095007

[18] Chlingaryan A, Sukkarieh S, Whelan B. Machine learning approaches for crop yield prediction and nitrogen status estimation in precision agriculture: A review. Computers and Electronics in Agriculture. 2018;**151**:61-69

[19] Rejeb A, Abdollahi A, Rejeb K, Treiblmaier H. Drones in agriculture: A review and bibliometric analysis. Computers and Electronics in Agriculture. 2022;**198**:107017

[20] Gerhards R, Risser P, Spaeth M, Saile M, Peteinatos G. A comparison of seven innovative robotic weeding systems and reference herbicide strategies in sugar beet (Beta vulgaris subsp. vulgaris L.) and rapeseed (Brassica napus L.). Weed Research. 2023;**64**(1):42-53. DOI: 10.1111/wre.12603

[21] Droukas L, Doulgeri Z, Tsakiridis NL, Triantafyllou D, Kleitsiotis I, Mariolis I, et al. A survey of robotic harvesting systems and enabling technologies. Journal of Intelligent & Robotic Systems. 2023;**107**(2):21

[22] Gonzalez-de-Soto M, Emmi L, Perez-Ruiz M, Aguera J, Gonzalez-de-Santos P. Autonomous systems for precise spraying–evaluation of a robotised patch sprayer. Biosystems Engineering. 2016;**146**:165-182

[23] Liu J, Cai H, Chen S, Pi J, Zhao L. A review on soil nitrogen sensing technologies: Challenges, Progress and perspectives. Agriculture; 2023;**13**(4):743

[24] Artizzu XPB, Ribeiro BA, Guijarro M, Pajares G. Real-time image processing for crop/weed discrimination in maize fields. Computer and Electronics in Agriculture. 2011;**75**:337-346. DOI: 10.1016/j.compag.2010.12.011

[25] Milioto A, Lottes P, Stachniss C. Real-time semantic segmentation of crop and weed for precision agriculture robots leveraging background knowledge in CNNs. In: 2018 IEEE International Conference on Robotics and Automation (ICRA). Brisbane, Australia: IEEE; 2018. pp. 2229-2235

[26] Kulkarni AH, Patil A. Applying image processing techniques to detect plant diseases. International Journal of Modern Engineering Research. 2012;**2**(5):3661-3664

[27] Liakos KG, Busato P, Moshou D, Pearson S, Bochtis D. Machine learning in agriculture. A review. Sensors. 2018;**18**(8):2674

[28] Sishodia RP, Ray RL, Singh SK. Applications of remote sensing in precision agriculture: A review. Remote Sensing. 2020;**12**(19):3136. DOI: 10.3390/rs12193136

[29] Hellerstein D, Vilorio D. Agricultural resources and environmental indicators. In: EIB- 208. Washington, D.C., USA: U.S. Department of Agriculture, Economic Research Service; May 2019

[30] Tsouros DC, Bibi S, Sarigiannidis PG. A review on UAV-based applications

for precision agriculture. Information. 2019;**10**(11):349

[31] Escolà A, Martínez-Casasnovas JA, Rufat J, Arnó J, Arbonés A, Sebé F, et al. Mobile terrestrial laser scanner applications in precision fruticulture/horticulture and tools to extract information from canopy point clouds. Precision Agriculture. 2017;**18**:111-132

[32] Heege HJ. Precision in guidance of farm machinery. Precision in crop farming: Site specific concepts and sensing methods. Applications and Results. 2013:35-50

[33] Serrano L, Penuelas J, Ustin SL. Remote sensing of nitrogen and lignin in Mediterranean vegetation from AVIRIS data: Decomposing biochemical from structural signals. Remote Sensing of Environment. 2002;**81**(2-3):355-364

[34] Jamali M, Soufizadeh S, Yeganeh B, Emam Y. Wheat leaf traits monitoring based on machine learning algorithms and high-resolution satellite imagery. Ecological Informatics. 2023;**74**:101967

[35] Gitelson AA. Wide dynamic range vegetation index for remote quantification of biophysical characteristics of vegetation. Journal of Plant Physiology. 2004;**161**(2):165-173

[36] Senbayram M, Chen R, Wienforth B, Herrmann A, Kage H, Mühling KH, et al. Emission of $N_2O$ from biogas crop production systems in northern Germany. Bioenergy Research. 2014;**7**:1223-1236

[37] Bramley RG, Lawes RA, Cook SE. Spatially distributed experimentation: Tools for the optimization of targeted management. In: Precision Agriculture for Sustainability and Environmental Protection. London: Routledge; 2013. pp. 205-218

[38] Colaço AF, Bramley RG. Do crop sensors promote improved nitrogen management in grain crops? Field Crops Research. 2018;**218**:126-140

[39] Biermacher JT, Epplin FM, Brorsen BW, Solie JB, Raun WR. Economic feasibility of site-specific optical sensing for managing nitrogen fertilizer for growing wheat. Precision Agriculture. 2009;**10**:213-230

[40] Boyer CN, Wade Brorsen B, Solie JB, Raun WR. Profitability of variable rate nitrogen application in wheat production. Precision Agriculture. 2011;**12**:473-487

[41] Roberts DC, Brorsen BW, Solie JB, Raun WR. The effect of parameter uncertainty on whole-field nitrogen recommendations from nitrogen-rich strips and ramped strips in winter wheat. Agricultural Systems. 2011;**104**(4):307-314

[42] Hochreiter S, Schmidhuber J. Long short-term memory. Neural Computation. 1997;**9**(8):1735-1780

[43] Elavarasan D, Vincent PD. Crop yield prediction using deep reinforcement learning model for sustainable agrarian applications. IEEE Access. 2020;**8**:86886-86901

[44] Girshick R, Donahue J, Darrell T, Malik J. Rich feature hierarchies for accurate object detection and semantic segmentation. In: 2014 IEEE Conference on Computer Vision and Pattern Recognition, Columbus, OH, USA. 2014. pp. 580-587. DOI: 10.1109/CVPR.2014.81

[45] Girshick R. Fast r-cnn. In: Proceedings of the IEEE International Conference on Computer Vision. Santiago, Chile: IEEE; 2015. pp. 1440-1448. DOI: 10.1109/ICCV.2015.169

[46] Ren S, He K, Girshick R, Sun J. Faster R-CNN: Towards real-time object detection with region proposal networks. Advances in Neural Information Processing Systems. 2015;**28**:28. DOI: 10.48550/arXiv.1506.01497

[47] Redmon J, Divvala S, Girshick R, Farhadi A. You only look once: Unified, real-time object detection. In: Proceedings of the IEEE Conference on Computer Vision and Pattern Recognition. Las Vegas, NV, USA: IEEE; 2016. pp. 779-788

[48] Wang S. Research towards YOLO-series algorithms: Comparison and analysis of object detection models for real-time UAV applications. In: Journal of Physics: Conference Series. Vol. 1948, no. 1. Bristol, England: IOP Publishing; 2021. p. 012021

[49] Liu W, Anguelov D, Erhan D, Szegedy C, Reed S, Fu CY, et al. SSD: Single shot multibox detector. In: Computer Vision–ECCV 2016: 14th European Conference, Amsterdam, the Netherlands, October 11-14, 2016, Proceedings, Part I 14. Cham, Switzerland: Springer International Publishing; 2016. pp. 21-37

[50] Divya R, Peter JD. Smart healthcare system-a brain-like computing approach for analyzing the performance of detectron2 and PoseNet models for anomalous action detection in aged people with movement impairments. Complex & Intelligent Systems. 2022;**8**(4):3021-3040

[51] Van Klompenburg T, Kassahun A, Catal C. Crop yield prediction using machine learning: A systematic literature review. Computers and Electronics in Agriculture. 2020;**177**:105709

[52] Chen CJ, Huang YY, Li YS, Chen YC, Chang CY, Huang YM. Identification of fruit tree pests with deep learning on embedded drone to achieve accurate pesticide spraying. IEEE Access. 2021;**9**:21986-21997

Chapter 8

# Nanoparticles in the Field: Sowing Innovation to Harvest a Sustainable Future

*Eliete A. Alvin, Wesley S.M. Ribeiro, Anna V.B. Borges, Rodrigo C. Rosa, Marcos V. Silva, Nilvanira D. Tebaldi and Anielle Christine A. Silva*

**Abstract**

The incorporation of nanoparticles into sensors or with antimicrobial and fertilizer properties in agriculture signifies a paradigm shift toward accuracy and sustainability. This chapter shows the numerous uses of nanoparticles and nanoparticle-based sensors in agribusiness. Their innovative contribution to promoting eco-friendly practices is examined. A bounty of advancements that maximize yields and contribute to a sustainable agricultural future is promised by these bactericidal nanoparticles, sensor technologies, and enhanced fertilizers, which sow the seeds of creativity. The discussion explores the potential of nanoparticles to transform farming practices, diminish environmental harm, and cultivate a progressive, eco-conscious farming landscape. Nanoparticle-based sensors provide data for informed decision-making, bactericidal nanoparticles protect crops from harmful pathogens, and nanotechnology enhances fertilizers for nutrient delivery and plant uptake.

**Keywords:** nanoparticles, bactericidal effects, sensor technologies, synergism, agriculture

## 1. Introduction

The incorporation of nanoparticles into agricultural practices represents a transformative paradigm shift, providing innovative solutions across multiple dimensions of contemporary farming [1, 2]. This encompasses controlled fertilizer release, crop protection, advanced agricultural sensing technologies, optimization of agrochemical delivery, improved water use efficiency, enhanced plant nutrition, and soil remediation [3]. Positioned as a promising remedy for more effective and environmentally sustainable agricultural practices, nanoparticles contribute to increased crop productivity, conservation of resources, and diminished environmental impacts.

Nanoparticles play a pivotal role in tackling key challenges and enhancing efficiency within agriculture. They facilitate controlled fertilizer release, ensuring optimal nutrient absorption, minimizing waste, and optimizing nutrient utilization [4, 5]. This transformative influence in agriculture extends to crop protection, utilizing nanoparticles with bactericidal and fungicidal properties to establish robust shields against pathogens, surpassing conventional methods for sustainable yield protection [6].

In agricultural sensing technologies, nanoparticles seamlessly integrate into state-of-the-art sensors, enabling real-time monitoring of environmental variables such as soil moisture, pH, nutrient concentration, and pest presence. This provides farmers with precise data for informed decision-making, raising the standards of agricultural management [7, 8]. Agrochemical delivery utilizing nanoparticles redefines the landscape by ensuring controlled and targeted release mechanisms, optimizing pesticide and herbicide effectiveness while minimizing environmental impacts [9, 10].

Nanoparticles significantly contribute to water conservation by improving soil water retention, reducing the need for frequent irrigation, and promoting efficient water resource utilization in agriculture [11, 12]. The integration of nanoparticles into plant nutrition products marks a paradigm shift, optimizing nutrient absorption and fostering healthier crops in an environmentally conscious manner [12].

In the remediation of contaminated soils, nanoparticles play a crucial role in absorbing and removing contaminants, offering a promising avenue for recovering degraded areas and contributing to sustainable soil restoration [13, 14]. As the subsequent sections unfold, each application will be comprehensively examined, elucidating nanoparticles' intricate role in steering agriculture toward a more sustainable, efficient, and environmentally sensitive future.

## 2. Nanoparticles with innovative properties in agriculture

The utilization of nanoparticles to achieve controlled fertilizer release holds immense potential in agriculture, offering benefits such as gradual nutrient discharge, improved plant absorption, and reduced fertilizer wastage, ultimately enhancing nutritional efficiency [15, 16]. This method is crucial in mitigating nutrient leaching, contributing to sustainable agricultural practices and environmental conservation. However, economic challenges may emerge due to initial higher production costs. Key technologies, such as mesoporous silica nanoparticles and functionalized carbon nanotubes, are instrumental in realizing optimal outcomes for controlled fertilizer release in agriculture [17, 18].

Nanoparticles (NPs) possessing bactericidal and fungicidal properties offer an efficient alternative for crop protection, diminishing reliance on conventional pesticides and reducing environmental contamination. However, challenges include potential resistance to development and intricate production processes, posing obstacles to widespread adoption [19]. Concerns regarding soil accumulation and residual effects on plant health and the environment underscore the necessity for thorough evaluations in the responsible application of nanoparticles. Economic feasibility and advancements in cost-effective production methods are critical considerations for widespread adoption in agriculture.

In economic considerations, the development and production of nanoparticle-based nutritional solutions may incur higher costs, necessitating a delicate balance between innovation and practicality for accessibility and widespread adoption in agriculture. The journey toward nanoparticle-enhanced plant nutrition entails promises, challenges, and a commitment to responsible and sustainable practices [20]. As nanoparticles reshape agricultural practices and address soil remediation challenges, continuous research and understanding will propel innovative and sustainable solutions for efficient and environmentally conscious agriculture [21].

**Figure 1** illustrates a summary of previous topics on the advantages of nanoparticles in plants, soil, seeds, and water.

## 2.1 Metallic nanoparticles

In agriculture, metal oxide nanoparticles like zinc oxide (ZnO), copper oxide (CuO), and titanium dioxide ($TiO_2$) serve multiple purposes, including controlled fertilizer release, crop protection, and soil remediation. Silver and Cu nanoparticles exhibit powerful antimicrobial properties, positioning them as eco-friendly alternatives to traditional pesticides. Including zeolite and $TiO_2$ nanoparticles in soil remediation effectively eliminates contaminants, contributing to rejuvenation. However, challenges such as environmental accumulation and economic feasibility require careful consideration [22, 23].

Notably, ZnO nanoparticles play a crucial role in agriculture, positively impacting crop nutrient delivery, seed germination, and stress tolerance. Tebaldi et al. demonstrated the potential of ZnO nanocrystals doped with several metals (Ag, Au, Mn, Ni, Cu, K, Mo, Cu, Fe, Cl) in controlling bacterial diseases, underscores their significance [24–27]. The doping process consists of incorporating elements from the crystalline structure of crystals, enhancing various applications. In the case of the tests, the incorporated ions enhanced the bactericidal effect of the ZnO nanoparticles. Addressing environmental concerns and toxicity is crucial for their responsible agricultural application [28].

Silver (Ag) and copper (Cu) nanoparticles, known for their potent antimicrobial properties, play a crucial role in crop protection, extending their applications beyond agriculture. Studies highlight their efficacy in controlling human pathogens and microorganisms in plants and seeds [29]. The positive impact of Ag NPs on spot diseases, seed dormancy, and seedling development underscores their potential [30, 31].

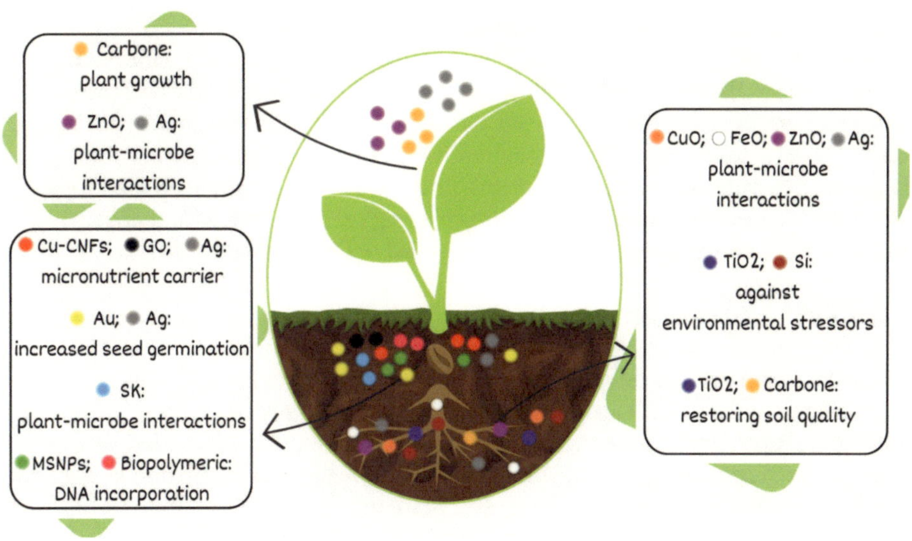


**Figure 1.**
*Types of nanoparticles and their applications.*

Research on the phytotoxicity of metallic nanoparticles, including Cu, CuO, Ag, and ZnO, emphasizes nuanced sensitivity levels, highlighting the need for regulation in their application [25, 26]. In soil remediation, zeolite and $TiO_2$ nanoparticles significantly remove contaminants and restore soil quality [32].

Iron nanoparticles (Fe NPs) are focal points in agricultural research, optimizing nutrient absorption and having diverse water and soil decontamination applications. Careful consideration of application rates and soil interactions is crucial for their positive impact [33].

The broader examination of metallic nanoparticles emphasizes the need for careful regulation and understanding their effects on plant germination [31]. Applying various nanoparticles at different plant development stages offers positive and negative outcomes, necessitating a nuanced approach. Overall, nanoparticle integration holds significant promise in advancing sustainable and effective agricultural practices.

### 2.2 Quantum dots

Quantum dots (QDs), nanoscale particles with distinct optical and electronic properties, have gained attention in agriculture for their potential applications. Their physicochemical characteristics, including high stability, water solubility, biocompatibility, adjustable surface functionalities, and low toxicity, make them promising tools for precise imaging and monitoring crops at the molecular level. This enables visualization of plant physiological processes, nutrient uptake, and stress responses, providing valuable insights into plant health and facilitating targeted interventions [34–36].

Environmental considerations regarding the fate of QDs in soil, water, and plants are crucial [37–40]. A thorough investigation into the long-term impacts on soil microbial communities and ecosystem dynamics is essential for responsible use in agriculture. Additionally, understanding the potential toxicity of QDs to plants and other organisms is vital for their safe integration into the agroecosystem. Balancing the advantages of QDs in precision agriculture with environmental sustainability is critical for their responsible use in modern farming practices.

Research by Ritesh Banerjee et al. explores the cytotoxicity and genotoxicity effects of CdSe QDs on a model plant (*Allium cepa*), revealing oxidative stress-induced activation of antioxidant scavengers and enzymes at low concentrations of CdSe QDs [41]. Another study by Sanghamitra Majumdar investigates the influence of exposure to cadmium sulfide quantum dots (CdS-QDs) on soybean seedlings, indicating the role of peroxidases in extinguishing oxidative stress and observing mechanisms of stress tolerance in plants exposed to QD treatments [42].

### 2.3 Carbon-based nanoparticles

Carbon nanoparticles hold significant promise in improving soil quality and supporting plant health. They act as effective carriers for nutrients, pesticides, and other agricultural inputs, enhancing soil water retention, optimizing irrigation efficiency, and strengthening crop drought resistance. Additionally, carbon nanoparticles exhibit bio-stimulant properties that foster plant growth and development [43, 44].

A comprehensive study by Darshan Rudakiya et al. explores the production, potential, and prospects of carbon nanotubes (CNTs) in agriculture, emphasizing the need for a nuanced understanding of CNT-plant interactions and their effects on plant-associated microbes [45]. Despite their benefits, challenges related to biodegradability arise due to the unique structural and electrical properties of CNTs [46].

Multi-walled carbon nanotubes applied to Brassica and broccoli species demonstrate increased tolerance to salinity stress, attributed to the promotion of the nitric oxide gas signaling molecule and aquaporin transduction in the respective plants [47, 48]. CNTs, a subset of QDs, exhibit enhanced resistance to drought stress in peanut plant seedlings, showcasing the diverse applications of nanomaterials in agriculture [49]. The versatility of QDs is highlighted in studies illustrating their use as non-invasive probes for visualizing plant physiological processes, tracking nutrient uptake, and detecting stress responses [50].

In the realm of graphene-based materials, including reduced graphene oxide (rGO) and graphene oxide (GO), researchers explore their potential applications in agriculture due to their unique physicochemical properties. Graphene's versatility in enhancing nutrient supply to the soil, promoting plant growth, and expanding crop yields is supported by agricultural research [51]. Studies reveal that adding graphene to the soil enhances nutrient retention, improving seed germination capacity and strengthening plants. Moreover, graphene derivatives improve cation exchange capacity and water retention in the soil, promoting better nutrient availability for plants [52]. Zhao et al. demonstrate the efficacy of graphene oxide as a water retention agent in soil, significantly increasing drought stress tolerance in soybean plants [53, 54].

Research on graphene derivatives elucidates their positive impacts on seed germination, root development, and photosynthetic efficiency. These improvements are attributed to the modulation of hormone levels, enzyme activities, and gene expression, showcasing graphene's potential in cultivating healthier and more robust crops [53]. Furthermore, graphene derivatives exhibit antimicrobial properties, effectively controlling pests and diseases in plants and seeds. These graphene-based nanomaterials offer a sustainable alternative to traditional pesticides, and their controlled-release formulations present a viable strategy for managing pathogen proliferation without causing environmental harm [55].

### 2.4 Silicon and selenium nanoparticles

Silicon nanoparticles have emerged as pivotal contributors to enhancing plants' resilience against stress, a pivotal aspect of sustainable agriculture. Experimental evidence highlights the positive impact of silicon nanoparticles on barley plants, showcasing not only enhanced recovery from drought stress but also significant increases in chlorophyll content (up to 17%) and shoot biomass (up to 27%) compared to control plants [56, 57]. The multifaceted influence of silicon nanoparticles on various physiological aspects underscores their potential as stress-alleviating agents in plant systems.

In a comprehensive study investigating the effects of potassium nanosilica (PNS) on corn plants facing drought stress, intriguing findings emerged under varying stress levels [58]. Drought stress led to a reduction in concentrations of essential elements in both aerial parts and seeds. Notably, the decrease in elemental concentrations coincided with an elevation in nitrogen concentration in seeds and an increase in potassium (K) concentration in aerial parts under water stress conditions. This emphasizes the dynamic role of potassium nanosilica in mitigating the adverse effects of drought stress on nutrient uptake and assimilation, contributing to an enhanced adaptive response in corn plants.

Skalický et al. observed that silicon nanoparticles (SiNPs) enhance the stress tolerance of plants against various environmental pressures, serving as a non-toxic and efficient alternative for controlling plant diseases [59].

Selenium nanoparticles (SeNPs) have also gained attention in agriculture for their potential benefits in enhancing plant growth and development. Studies indicate that SeNPs improve nutrient absorption, increase selenium utilization efficiency, and provide protection against abiotic stresses, contributing to enhanced crop resilience. These nanoparticles exhibit antioxidant properties, mitigating oxidative stress in plants and promoting overall plant health [60].

The ability of SeNPs to biofortify food crops with selenium is noteworthy, addressing selenium deficiencies in human diets. However, careful consideration is essential due to concerns such as precise dosage control to avoid toxicity, potential environmental impacts, and the safety of incorporating nanomaterials into the food chain. Furthermore, the long-term effects of selenium nanoparticles on soil microbiota and ecosystems necessitate thorough investigation to ensure sustainable and responsible agricultural practices [61, 62].

Ongoing research aims to elucidate the nuanced benefits and challenges of using selenium nanoparticles in agriculture, facilitating their informed and judicious application. In a study by Jingfu et al., selenium nanoparticles offered diverse plant advantages, including improved nutrient absorption, enhanced selenium utilization efficiency, protection against abiotic stresses, promotion of growth, and potential for food biofortification. These nanoparticles can also reduce dependence on conventional fertilizers and hold promise in soil phytoremediation. However, careful consideration is crucial in addressing environmental and safety concerns associated with using materials at the nanoscale [63].

### 2.5 Biopolymeric nanoparticles

Biopolymeric nanoparticles have emerged as a promising agricultural solution, leveraging their natural origin and biodegradable properties. These nanoparticles are derived from substances like proteins and polysaccharides and offer a sustainable alternative to synthetic materials. Their notable application lies in enabling the controlled release of nutrients, optimizing plant absorption, and minimizing fertilizer wastage. They prove versatile in crop protection, acting as carriers for bioactive agents to gradually and precisely enhance resistance against pathogens and pests.

Moreover, these nanoparticles stimulate plant growth by encapsulating biostimulants, promoting overall crop health and productivity. Despite their considerable potential, challenges related to synthesis complexity and potential environmental impacts necessitate ongoing research to ensure their responsible and effective integration into agricultural practices [64, 65].

In sustainable agriculture, the adoption of biopolymeric nanoparticles holds promise due to their inherent biodegradability and eco-friendly attributes [66, 67]. These nanoparticles are sourced from proteins and polysaccharides and offer a green alternative to conventional synthetic materials. Their efficacy in nutrient management is particularly noteworthy, allowing for a controlled and sustained release of essential elements. This enhances the efficiency of nutrient absorption by plants while mitigating the negative environmental effects associated with excessive fertilizer application [68].

Furthermore, the application of biopolymeric nanoparticles in crop protection represents a sophisticated approach. Here, the nanoparticles act as carriers for bioactive compounds, facilitating targeted and gradual release to reinforce plants against pathogens and pests. Additionally, the encapsulation of bio-stimulants within these nanoparticles contributes to crops' overall well-being and productivity [69–71].

Despite these promising attributes, challenges related to the complex synthesis of biopolymeric nanoparticles and potential environmental impacts underscore the need for ongoing research. This ensures a nuanced understanding of their role and impact on agricultural systems [65].

### 2.6 Nanoparticles to agricultural sensor

Integrating nanoparticles into agricultural sensors is paramount in enhancing the precision and efficacy of detection methodologies within the agricultural domain. This strategic amalgamation brings forth many benefits, contributing to the advancement of agricultural practices [72].

The addition of nanoparticles to sensors contributes unprecedented precision in detecting various environmental variables critical to agricultural processes. These nanoparticles, with their unique properties, enable sensors to capture nuanced data, providing a granular understanding of factors such as soil moisture, pH levels, nutrient concentration, and the presence of pests [73, 74].

Nanoparticle-infused sensors empower real-time monitoring of agricultural conditions. This capability allows for timely and accurate data collection, enabling farmers and agricultural practitioners to make informed decisions promptly. Real-time monitoring is crucial in addressing dynamic changes in the agricultural environment [75, 76].

Nanoparticles enhance the sensitivity of agricultural sensors, ensuring their responsiveness to subtle changes in the surroundings. This heightened sensitivity is instrumental in detecting early signs of potential issues, such as variations in nutrient levels or the onset of pest infestations, enabling proactive intervention [77, 78].

The incorporation of nanoparticles broadens the scope of detection capabilities within agricultural sensors. These sensors can simultaneously measure multiple parameters, offering a comprehensive assessment of the agricultural landscape. This multifunctional approach is invaluable for a holistic understanding of the factors influencing crop health and productivity [79].

**Figure 2** illustrates the operating process of a biosensor, as well as the main regions of interest, such as, for example, the bioreceptor layer, in addition to illustrating the main nanoparticles that can be used to improve the sensor's analyte response.

The precise and real-time data of nanoparticle-enhanced sensors facilitates improved agricultural decision-making. Farmers can make data-driven choices regarding irrigation schedules, fertilizer application, and pest control measures, leading to optimized resource utilization and increased efficiency [80].

By enhancing the accuracy and efficiency of detection, nanoparticle-infused sensors contribute to sustainable agricultural practices. This sustainability is manifested through the judicious use of resources, reduced environmental impact, and promotion of eco-friendly farming methodologies.

Incorporating nanoparticles into agricultural sensors represents a pivotal advancement in detection capabilities. The resulting enhancement in precision, real-time monitoring, sensitivity, and multifunctionality contributes to informed decision-making and fosters sustainability in agricultural practices. As technology continues to evolve, the synergy between nanoparticles and agricultural sensors is poised to play a central role in shaping the future of precision agriculture [81–83].

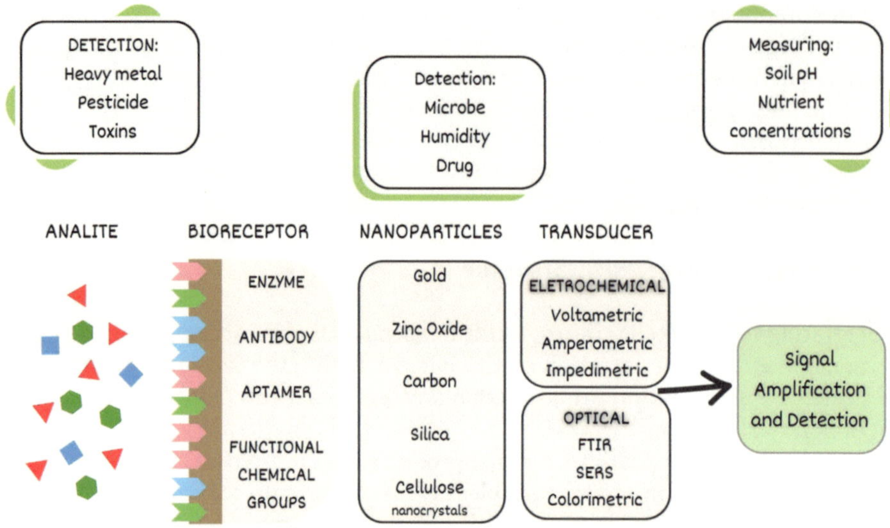


**Figure 2.**
*The figure illustrates the operating process of a biosensor and the types of nanoparticles used.*

Various nanoparticles play pivotal roles in constructing advanced agricultural sensors, each contributing unique attributes to enhance precision monitoring capabilities. Gold (Au), zinc oxide (ZnO), carbon nanotubes, and silica (Si) stand out as critical materials, each serving specific functions in soil condition assessment and environmental monitoring [84–86].

Gold nanoparticles are chosen for their outstanding conductivity and stability, making them ideal for sensor applications. In agricultural sensors, these nanoparticles efficiently capture and transmit data related to soil conditions. Their electrical properties enable precise signal transmission, allowing for high-quality data acquisition and comprehensive soil analysis.

ZnO nanoparticles exhibit enhanced sensing capabilities, particularly in measuring soil pH and nutrient concentrations. Silva et al. demonstrated the interaction of ZnO nanoparticles with glyphosate molecules for infrared detection and how pH changes the response [87, 88].

Renowned for their exceptional strength and electrical conductivity, carbon nanomaterials are crucial in sensors that monitor various environmental variables. Their robustness and electrical properties make them well-suited for continuous real-time data collection in precision agriculture, facilitating timely decision-making in response to changing conditions.

Silica nanoparticles are known for their versatility and ease of functionalization. Their ability to be easily functionalized in sensor applications contributes to developing sensors capable of simultaneously detecting multiple parameters. The versatility of silica nanoparticles enhances the adaptability and efficiency of sensors in diverse agricultural settings, allowing for comprehensive and multifaceted data collection [89].

In agricultural sensing technology, the strategic integration of these nanoparticles underscores the pursuit of precision and efficiency. The unique properties of gold, ZnO, carbon nanotubes, and silica empower sensors to provide accurate, real-time data, advancing precision agriculture and sustainable farming practices. As research continues, the potential for further innovation and optimization in nanoparticle-based sensors remains promising.

Yao Yao et al. proposed developing a new approach to increase the humidity sensitivity of quartz crystal microbalance (QCM) humidity sensors. By the asymmetric treatment of the QCM electrode, the fabricated QCM humidity sensor produces additional frequency changes associated with the change of the dielectric constant of the sensing material in addition to the bulk sensitivity. Renewable cellulose nanocrystals (CNCs) as humidity-sensing material were deposited on the sensing electrode of QCM. The results proved that the humidity sensitivity of the QCM humidity sensor based on CNCs can be effectively improved by the asymmetric treatment of the QCM electrode structure [90].

## 3. Conclusion

The integration of nanoparticles in agriculture, mainly through innovative sensors, bactericidal agents, and enhanced fertilizers, signifies a groundbreaking shift toward precision and sustainability. This chapter emphasizes the transformative potential of nanoparticles in fostering sustainable agricultural practices. By promoting innovation, nanoparticles offer a promising avenue for advancements that optimize crop yields and contribute to a more sustainable agricultural future. Their ability to enhance agricultural efficiency, minimize environmental impact, and facilitate real-time data acquisition underscores their significance in redefining agricultural practices for a harmonious balance between productivity and environmental responsibility.

## Acknowledgements

This work was supported by grants from CNPq, CAPES, FAPEAL, RENORBIO, and FAPEMIG.

## Conflict of interest

The authors declare no conflict of interest.

## Author details

Eliete A. Alvin[1,2], Wesley S.M. Ribeiro[1], Anna V.B. Borges[3], Rodrigo C. Rosa[4], Marcos V. Silva[3], Nilvanira D. Tebaldi[5] and Anielle Christine A. Silva[1,2]*

1 Strategic Materials Laboratory, Physics Institute, Federal University of Alagoas, Maceió, Alagoas, Brazil

2 Northeast Biotechnology Network, Federal University of Alagoas, Maceió, AL, Brazil

3 Department of Microbiology, Immunology and Parasitology, Institute of Biological and Natural Sciences, Federal University of Triângulo Mineiro, Uberaba, Minas Gerais, Brazil

4 Department of Structural Biology, Institute of Biological and Natural Sciences, Federal University of Triângulo Mineiro, Uberaba, Minas Gerais, Brazil

5 Laboratory of Plant Bacteriology, Institute of Agricultural Sciences, Federal University of Uberlândia, Uberlândia, Minas Gerais, Brazil

*Address all correspondence to: acalmeida@fis.ufal.br

## References

[1] Ghouri F, Shahid MJ, Zhong M, Zia MA, Alomrani SO, Liu J, et al. Alleviated lead toxicity in rice plant by co-augmented action of genome doubling and TiO2 nanoparticles on gene expression, cytological and physiological changes. Science of the Total Environment [Internet]. 2024;**911**:168709. Available from: https://www.sciencedirect.com/science/article/pii/S0048969723073370

[2] Naseer M, Yang YM, Zhu Y, Zhao L, Cao J, Wang S, et al. Nano-iron and AM fungi inoculation in dryland wheat field: A sustainable alternative to plastic film mulching. Field Crops Research [Internet]. 2024;**306**:109208. Available from: https://www.sciencedirect.com/science/article/pii/S037842902300401X

[3] Sharma P, Sangwan S, Mehta S. Chapter 3 - Emerging role of phosphate nanoparticles in agriculture practices. In: Husen A, editor. Engineered Nanomaterials for Sustainable Agricultural Production, Soil Improvement and Stress Management [Internet], Plant Biology, Sustainability and Climate Change. Academic Press; 2023. pp. 71-97. Available from: https://www. sciencedirect.com/science/article/pii/ B9780323919333000088 [Accessed: Jan 9, 2024]

[4] Rana A, Pathak S, Kumar K, Kumari A, Chopra S, Kumar M, et al. Multifaceted properties of TiO2 nanoparticles synthesized via Mangifera indica and Azadirachta indica plant extracts: Antimicrobial, antioxidant, and non-linear optical investigations for sustainable agriculture applications. In: Materials Advances. RSC Publishing; 2024. DOI: 10.1039/D3MA00414G [Internet]. Available from: https://pubs.rsc.org/en/content/articlehtml/2024/ma/d3ma00414g [Accessed: Jan 9, 2024]

[5] Goyal U, Chaubey KK, Pandey SD, Verma DK, Bachheti A, Bachheti RK. Role of gold nanoparticles in plant protection against pathogen. In: Bachheti RK, Bachheti A, Husen A, editors. Metal and Metal-Oxide Based Nanomaterials: Synthesis, Agricultural, Biomedical and Environmental Interventions [Internet], Smart Nanomaterials Technology. Singapore: Springer Nature; 2024. pp. 201-211. DOI: 10.1007/978-981-99-7673-7_10 [Accessed: Jan 9, 2024]

[6] Bhaskar R, Pandey SP, Kumar U, Kim H, Jayakodi SK, Gupta MK, et al. Nanobionics for sustainable crop production: Recent development to regulate plant growth and protection strategies from pests. OpenNano. 2024;**15**:100198. Available from: https://www. sciencedirect.com/science/article/pii/ S2352952023000774 [Accessed: Jan 9, 2024]

[7] Abdali M, Ghasemi F, Seyed Hosseini HM, Mahdavi V. Different sized gold nanoparticles for array-based sensing of pesticides and its application for strawberry pollution monitoring. Talanta [Internet]. 2024;**267**:125121. Available from: https://www.sciencedirect.com/science/article/pii/S003991402300872X

[8] Periasamy P, Gowtham G, Selvakumar B, Nikolova MP, Bououdina M, Supangat A, et al. Application of green graphene-based nanomaterials in agri-biotechnological sensors for surveillance and prevention of agricultural productivity. In: Graphene-Based Nanomaterials. 2024. pp. 109-120. Available from: https://www.taylorfrancis.com/chapters/edit/10.1201/9781003300540-11/

application-green-graphene-based-nanomaterials-agri-biotechnological-sensors-surveillance-prevention-agricultural-productivity-periasamy-gowtham-selvakumar-maria-plamenova-nikolova-mohamed-bououdina-azzuliani-supangat-atef-shenouda-devarajan

[9] Verma R, Saroop S. Chapter 11 - Impact of nanopesticides in the environment: Solutions, threats, and opportunities. In: Sharma A, Kumar V, Zheng B, editors. Pesticides in the Environment [Internet]. Elsevier; 2024. pp. 251-292. Available from: https://www.sciencedirect.com/science/article/pii/B9780323994279000100 [Accessed: Jan 9, 2024]

[10] Saryeddine L, Hadnutt J, Grélard A, Morvan E, Alies B, Buré C, et al. Design of light-responsive amphiphilic self-assemblies: A novel application of the photosensitive diazirine moiety. Journal of Colloid and Interface Science [Internet]. 2024;**653**:1792-1804. Available from: https://www.sciencedirect.com/science/article/pii/S0021979723019094

[11] Vyas TK, More B, Mehta MP. Chapter 5 - Improving stress resilience in plants by nanoparticles. In: Ahanger MA, Bhat JA, Ahmad P, John R, editors. Improving Stress Resilience in Plants [Internet]. Academic Press; 2024. pp. 73-96. Available from: https://www.sciencedirect.com/science/article/pii/B9780443189272000236 [Accessed: Jan 9, 2024]

[12] Leal FDS, Santos Neto H, Pinheiro ICL, Oliveira JM, Pozza AAA, Pozza E. A copper and silver nanoparticles control coffee rust: Decrease the quantity of sprayed active ingredients and is an alternative for sustainable coffee production. European Journal of Plant Pathology [Internet]. 2024;**168**(1):39-51. Available from: https://www.taylorfrancis.com/chapters/edit/10.1201/9781003300540-11/

application-green-graphene-based-nanomaterials-agri-biotechnological-sensors-surveillance-prevention-agricultural-productivity-periasamy-gowtham-selvakumar-maria-plamenova-nikolova-mohamed-bououdina-azzuliani-supangat-atef-shenouda-devarajan

[13] Pal R, Kumar L, Anand S, Bharadvaja N. Environmental pollutants remediation using Phyto-nanoparticles: An overview on synthesis, characterization, and remediation potential. In: Shah MP, Bharadvaja N, Kumar L, editors. Biogenic Nanomaterials for Environmental Sustainability: Principles, Practices, and Opportunities [Internet], Environmental Science and Engineering. Cham: Springer International Publishing; 2024 [Accessed: Jan 9, 2024]. pp. 111-145. DOI: 10.1007/978-3-031-45956-6_5

[14] Sirajuddin A, MS. Convetional bioremediation strategies versus green nanoremediation. In: Green Nanoremediation: Sustainable Management of Environmental Pollution. Cham: Sringer International Publishing; 2023. pp. 333-356. Available from: https://link.springer.com/chapter/10.1007/978-3-031-30558-0_15 [Accessed: Jan 9, 2024]

[15] Affrald RJ. Large-scale production and application of graphene oxide nanoparticles to meet agriculture needs. Current Applied Science and Technology [Internet]. 2024;**24**(1):e0254564. DOI: 10.55003/cast.2023.254564. Available from: https://li01.tci-thaijo.org/index.php/cast/article/view/254564

[16] Sharma P, Aggarwal K, Pal AK, Chaubey KK, Sharma S, Verma DK, et al. Role of zinc nanoparticles for the Management of Post-harvest Diseases. In: Bachheti RK, Bachheti A, Husen A, editors. Metal and Metal-Oxide Based Nanomaterials: Synthesis, Agricultural,

Biomedical and Environmental Interventions [Internet], Smart Nanomaterials Technology. Singapore: Springer Nature; 2024. pp. 95-109. DOI: 10.1007/978-981-99-7673-7_5 [Accessed: Jan 9, 2024]

[17] Karnwal A, et al. Unveiling the potential of bioinoculants and nanoparticles in sustainable agriculture for enhanced plant growth and food security. BioMed Research International. [Internet]. 29 Nov 2023;**2023**:6911851. DOI: 10.1155/2023/6911851 [Accessed: Jan 9, 2024]

[18] Al-Khayri JM, Rashmi R, Surya Ulhas R, Sudheer WN, Banadka A, Nagella P, et al. The role of nanoparticles in response of plants to abiotic stress at physiological, biochemical, and molecular levels. Plants. 2023;**12**(2):292. Available from: https://doi.org/10.3390/plants1202029 Plants – Free full-text – The role [Accessed: Jan 9, 2024]

[19] Salem SS, Mekky AE. Biogenic nanomaterials: Synthesis, characterization, and applications. In: Shah MP, Bharadvaja N, Kumar L, editors. Biogenic Nanomaterials for Environmental Sustainability: Principles, Practices, and Opportunities [Internet], Environmental Science and Engineering. Cham: Springer International Publishing; 2024. pp. 13-43. DOI: 10.1007/978-3-031-45956-6_2 [Accessed: Jan 9, 2024]

[20] Kust G, Andreeva O, Cowie A. Land degradation & development. In: Environmental & Soil Science Journal. Vol. 195. Wiley Online Journal [Internet]; 2017. pp. 16-24. DOI: 10.1016/j.jenvman.2016.10.043 [Accessed: Jan 9, 2024]

[21] ul Din K, Naeem MS, Zulifqar U, Albadrani GM, Waraich EA, Hussain S. Nanoparticles based biofortification in food crops: Overview, implications, and prospects. In: Hasanuzzaman M, Tahir MS, Tanveer M, Shah AN, editors. Mineral Biofortification in Crop Plants for Ensuring Food Security [Internet]. Singapore: Springer Nature; 2023 [Accessed: Jan 9, 2024]. pp. 173-201. DOI: 10.1007/978-981-99-4090-5_8

[22] Gauba A, Hari SK, Ramamoorthy V, Vellasamy S, Govindan G, Valan AM. The versatility of green synthesized zinc oxide nanoparticles in sustainable agriculture: A review on metal-microbe interaction that rewards agriculture. Physiological and Molecular Plant Pathology [Internet]. 2023;**125**:102023. Available from: https://www.sciencedirect.com/science/article/pii/S0885576523000784

[23] Iqbal S, Hussain M, Sadiq S, Seleiman MF, Sarkhosh A, Chater JM, et al. Silicon nanoparticles confer hypoxia tolerance in citrus rootstocks by modulating antioxidant activities and carbohydrate metabolism. Heliyon [Internet]. 2024;**10**(1):e22960. Available from: https://linkinghub.elsevier.com/retrieve/pii/S240584402310168X

[24] Silva ACA, Alvin EA, dos Santos FRA, de Matos SLM, de Oliveira JM, Silva AS, et al. Doped semiconductor nanocrystals: Development and applications [Internet]. In: Nanocrystals. IntechOpen; 2021. DOI: 10.5772/intechopen.96753

[25] Mamede MC, Mota RP, Silva ACA, Tebaldi ND. Nanoparticles in inhibiting *Pantoea ananatis* and to control maize white spot. Ciencia Rural [Internet]. 2021;**52**:e20210481. Available from: https://www.scielo.br/j/cr/a/bqm4S6YrG3Bqyh8YBj8SR8B/

[26] Fraga FS, Silva ACA, Dantas NO, Tebaldi ND, Luz JMQ. Doped zinc-oxide nanocrystals for the control of tomato bacterial spot and Xanthomonas gardneri in seeds. Tropical Plant Pathology [Internet]. 2021;**46**(4):406-413. DOI: 10.1007/s40858-021-00436-2

[27] Oliveira NS, Silva ACA, Tebaldi ND. Simonkolleite nanoparticles for seed treatment and control of tomato bacterial spot caused by *Xanthomonas hortorum* pv. *Gardneri*. Ciênc Agrotec. 2023;**47**:e000623. Available from: https://www.scielo.br/j/cagro/a/ZRvt6x8n76fnV7ghyfrFvjx/

[28] Tripathi GD, Javed Z, Dashora K. Toxicity of copper oxide nanoparticles on agriculturally important soil rhizobacteria bacillus megaterium. Emerging Contaminants. 2024;**10**(1):100280. Available from: https://www.sciencedirect.com/science/article/pii/S2405665023000781

[29] Mishra S, Singh BR, Singh A, Keswani C, Naqvi AH, Singh HB. Biofabricated silver nanoparticles act as a strong fungicide against Bipolaris sorokiniana causing spot blotch disease in wheat. PLoS One. 19 May 2014;**9**(5):e97881 [Internet]. Available from: https://journals.plos.org/plosone/article?id=10.1371/journal.pone.0097881

[30] Singh RP, Rai AR, Jotania RB, Chaudhary RG, Abdala A. Rationale and trends of applied nanotechnology. In: Biogenic Sustainable Nanotechnology. ScienceDirect [Internet]; 2022. pp. 337-389. Available from: https://www.sciencedirect.com/science/article/abs/pii/B9780323885355000081 [Accessed: Jan 9, 2024]

[31] Santás-Miguel V, Arias-Estévez M, Rodríguez-Seijo A, Arenas-Lago D. Use of metal nanoparticles in agriculture. A review on the effects on plant germination. Environmental Pollution. [Internet]. 2023;**334**:122222. Available from: https://www.sciencedirect.com/science/article/pii/S0269749123012241

[32] Abdel Latef AAH, Srivastava AK, El-sadek MSA, Kordrostami M, Tran LSP. Titanium dioxide nanoparticles improve growth and enhance tolerance of broad bean plants under saline soil conditions. Land Degradation & Development [Internet]. 2018;**29**(4):1065-1073. DOI: 10.1002/ldr.2780

[33] Yousaf N, Ishfaq M, Qureshi HA, Saleem A, Yang H, Sardar MF, et al. Characterization of root and foliar-applied iron oxide nanoparticles (α-Fe2O3, γ-Fe2O3, Fe3O4, and bulk Fe3O4) in improving maize (Zea mays L.) performance. Nanomaterials [Internet]. 2023;**13**(23):3036. Available from: https://www.mdpi.com/2079-4991/13/23/3036

[34] Gupta A, Mehta SK, Kunal K, Mukhopadhyay K, Singh S. 10 - quantum dots as promising nanomaterials in agriculture. In: Ghosh S, Thongmee S, Kumar A, editors. Agricultural Nanobiotechnology [Internet]. Woodhead Publishing Series in Food Science, Technology and Nutrition; 2022. pp. 243-296. Available from: https://www.sciencedirect.com/science/article/abs/pii/B978032391908100016X [Accessed: Jan 9, 2024]

[35] Lin X, Klenk R, Wang L, Köhler T, Albert J, Fiechter S, et al. 11.3% efficiency Cu(In,Ga)(S,Se)2 thin film solar cells via drop-on-demand inkjet printing. Energy & Environmental Science [Internet]. 2016;**9**(6):2037-2043. Available from: https://pubs.rsc.org/en/content/articlelanding/2016/ee/c6ee00587j

[36] Silva ACA, de Deus SLV, Silva MJB, Dantas NO. Highly stable luminescence of CdSe magic-sized quantum dots in HeLa cells. ScienceDirect Sensors and Actuators B: Chemical [Internet]. 2014;**191**:108-114. Available from: https://www.sciencedirect.com/science/article/abs/pii/S0925400513011040 [Accessed: Jan 17, 2024]

[37] Silva ACA, Freschi APP, Rodrigues CM, Matias BF, Maia LP, Goulart LR, et al. Biological analysis and

imaging applications of CdSe/CdSxSe1−x/CdS core–shell magic-sized quantum dot. Nanomedicine: Nanotechnology, Biology and Medicine [Internet]. 2016;**12**(5):1421-1430. Available from: https://www.sciencedirect.com/science/article/pii/S154996341600071X

[38] Goto TE, Lopes CC, Nader HB, Silva ACA, Dantas NO, Siqueira JR, et al. CdSe magic-sized quantum dots incorporated in biomembrane models at the air–water interface composed of components of tumorigenic and non-tumorigenic cells. Biochimica et Biophysica Acta (BBA) - Biomembranes [Internet]. 2016;**1858**(7, Part A):1533-1540. Available from: https://www.sciencedirect.com/science/article/pii/S0005273616301365

[39] Silva ACA, da Silva SW, Morais PC, Dantas NO. Shell thickness modulation in Ultrasmall CdSe/CdSxSe1–x/CdS Core/Shell quantum dots via 1-Thioglycerol. ACS Nano [Internet]. 2014;**8**(2):1913-1922. DOI: 10.1021/nn406478f

[40] Saraiva AL, Vieira TN, Notário AFO, Luiz JPM, Silva CR, Goulart LR, et al. CdSe magic-sized quantum dots attenuate reactive oxygen species generated by neutrophils and macrophages with implications in experimental arthritis. Nanomedicine: Nanotechnology, Biology and Medicine [Internet]. 2022;**42**:102539. Available from: https://www.sciencedirect.com/science/article/pii/S1549963422000259

[41] Banerjee R, Goswami P, Chakrabarti M, Chakraborty D, Mukherjee A, Mukherjee A. Cadmium selenide (CdSe) quantum dots cause genotoxicity and oxidative stress in Allium cepa plants. Mutation Research/Genetic Toxicology and Environmental Mutagenesis [Internet]. 2021;**865**:503338. Available from: https://www.sciencedirect.com/science/article/pii/S1383571821000292

[42] Majumdar S, Ma C, Villani M, Zuverza-Mena N, Pagano L, Huang Y, et al. Surface coating determines the response of soybean plants to cadmium sulfide quantum dots. NanoImpact. [Internet]. 2019;**14**:100151. Available from: https://www.sciencedirect.com/science/article/pii/S2452074818301848

[43] Nishshankage K, Fernandez AB, Pallewatta S, Buddhinie PKC, Vithanage M. Current trends in antimicrobial activities of carbon nanostructures: Potentiality and status of nanobiochar in comparison to carbon dots. Biochar [Internet]. 2024;**6**(1):2. DOI: 10.1007/s42773-023-00282-2

[44] Das NS, Pratim P. Role and application of graphene nanomaterial in crop improvement and the enhancement of productivity of crops. In: Graphene-Based Nanomaterials [Internet]. CRC Press; 2024. pp. 138-150. Avaliabe from: https://www.taylorfrancis.com/chapters/edit/10.1201/9781003300540-13/role-application-graphene-nanomaterial-crop-improvement-enhancement-productivity-crops-niraj-singh-pranjal-pratim-das

[45] Rudakiya D, Patel Y, Chhaya U, Gupte A. Carbon nanotubes in agriculture: Production, potential, and prospects. In: Panpatte DG, Jhala YK, editors. Nanotechnology for Agriculture: Advances for Sustainable Agriculture. [Internet]. Singapore: Springer; 2019. pp. 121-130. DOI: 10.1007/978-981-32-9370-0_8 [Accessed: Jan 9, 2024]

[46] Zhou J, Booker C, Li R, Zhou X, Sham TK, Sun X, et al. An electrochemical avenue to blue luminescent nanocrystals from multiwalled carbon nanotubes (MWCNTs). Journal of the American Chemical Society [Internet]. 2007;**129**(4):744-745. DOI: 10.1021/ja0669070

[47] Zhao G, Zhao Y, Lou W, Su J, Wei S, Yang X, et al. Nitrate Reductase-Dependent Nitric Oxide Is Crucial for Multi-Walled Carbon Nanotube-Induced Plant Tolerance against Salinity, Nanoscale. Vol. 11, no. 21. RSC Publishing [Internet]; 2019. pp. 10511-10523. DOI: 10.1039/C8NR10514F. Available from: https://pubs.rsc.org/en/content/articlelanding/2019/nr/c8nr10514f/unauth [Accessed: Jan 9, 2024]

[48] Martínez-Ballesta MC, Zapata L, Chalbi N, Carvajal M. Multiwalled carbon nanotubes enter broccoli cells enhancing growth and water uptake of plants exposed to salinity. Journal of Nanobiotechnology [Internet]. 2016;**14**(1):42. Available from: https://link.springer.com/article/10.1186/s12951-016-0199-4 [Accessed: Jan 9, 2024]

[49] Su LX, Ma XL, Zhao KK, Shen CL, Lou Q, Yin DM, et al. Carbon nanodots for enhancing the stress resistance of peanut plants. ACS Omega [Internet]. 2018;**3**(12):17770-17777. DOI: 10.1021/acsomega.8b02604

[50] Almeida Silva AC, Silva MJB, da Luz FAC, Silva DP, de Deus SLV, Oliveira DN. Controlling the cytotoxicity of CdSe magic-sized quantum dots as a function of surface defect density. Nano Letters [Internet]. 2014;**14**(9):5452-5457. DOI: 10.1021/nl5028028

[51] Das P, Khare P. Chapter 8 - agricultural waste-derived graphene and its derivatives: Opening the way for a sustainable environment. In: Hussain CM, Srivastav AK, Tiwary CS, Tour JM, editors. Graphene Extraction from Waste [Internet]. Woodhead Publishing Series in Electronic and Optical Materials. 2023. pp. 213-237. Available from: https://www.sciencedirect.com/science/article/pii/B9780323909143000024 [Accessed: Jan 9, 2024]

[52] Klionsky DJ, Abdel-Aziz AK, Abdelfatah S, Abdellatif M, Abdoli A, Abel S, et al. Guidelines for the use and interpretation of assays for monitoring autophagy (4th edition). Autophagy. 2021;**17**(1):1-382

[53] Zhang X, Cao H, Wang H, Zhao J, Gao K, Qiao J, et al. The effects of graphene-family nanomaterials on plant growth: A review. Nanomaterials [Internet]. 2022;**12**(6):936. Available from: https://www.mdpi.com/2079-4991/12/6/936

[54] Zhao L, Wang W, Fu X, Liu A, Cao J, Liu J. Graphene oxide, a novel nanomaterial as soil water retention agent, dramatically enhances drought stress tolerance in soybean plants. Frontiers in Plant Science [Internet]. 2022;**13**:810905. DOI: 10.3389/fpls.2022.810905

[55] Ray MK, Mishra AK, Mohanta YK, Mahanta S, Chakrabartty I, Kungwani NA, et al. Nanotechnology as a promising tool against Phytopathogens: A futuristic approach to agriculture. Agriculture [Internet]. 2023;**13**(9):1856. Available from: https://www.mdpi.com/2077-0472/13/9/1856

[56] Sarkar J, Mridha D, Sarkar J, Orasugh JT, Gangopadhyay B, Chattopadhyay D, et al. Synthesis of nanosilica from agricultural wastes and its multifaceted applications: A review. Biocatalysis and Agricultural Biotechnology [Internet]. 2021;**37**:102175. Available from: https://www.sciencedirect.com/science/article/pii/S1878818121002711

[57] Ghorbanpour M, Mohammadi H, Kariman K. Nanosilicon-based recovery of barley (Hordeum vulgare) plants subjected to drought stress. Environmental Science: Nano [Internet]. 2020;**7**(2):443-461. Available from: https://pubs.rsc.org/en/content/articlelanding/2020/en/c9en00973f

[58] Aqaei P, Weisany W, Diyanat M, Razmi J, Struik PC. Response of maize (Zea mays L.) to potassium nano-silica application under drought stress. Journal of Plant Nutrition. [Internet]. 2020;**43**(9):1205-1216. DOI: 10.1080/01904167.2020.1727508

[59] Mahawar L, Ramasamy KP, Suhel M, Prasad SM, Živčák M, Brestic M, et al. Silicon nanoparticles: Comprehensive review on biogenic synthesis and applications in agriculture. Environmental Research [Internet]. 2023;**232**:116292. Available from: https://www.sciencedirect.com/science/article/pii/S0013935123010964

[60] Bano I, Skalickova S, Sajjad H, Skladanka J, Horky P. Uses of selenium nanoparticles in the plant production. Agronomy. 2021;**11**(11):2229. Available from: https://www.mdpi.com/2073-4395/11/11/2229

[61] Zohra E, Ikram M, Omar A, Hussain M, Satti S, Raja N, et al. Potential applications of biogenic selenium nanoparticles in alleviating biotic and abiotic stresses in plants: A comprehensive insight on the mechanistic approach and future perspectives. Green Processing and Synthesis. 2021;**10**:456-475

[62] Ikram M, Javed B, Raja NI, Mashwani Z, ur R. Biomedical potential of plant-based selenium nanoparticles: A comprehensive review on therapeutic and mechanistic aspects. International Journal of Nanomedicine [Internet]. 2021;**16**:249. Available from: https://www.ncbi.nlm.nih.gov/pmc/articles/PMC7811472/

[63] Song J, Yu S, Yang R, Xiao J, Liu J. Opportunities for the use of selenium nanoparticles in agriculture. NanoImpact. [Internet]. 2023;**31**:100478. Available from: https://www.sciencedirect.com/science/article/pii/S2452074823000290

[64] Ramaprabha K, Venkat Kumar S, Saravanan P, Rajeshkannan R, Rajasimman M, Kamyab H, et al. Exploring the diverse applications of carbohydrate macromolecules in food, pharmaceutical, and environmental technologies. Environmental Research [Internet]. 2024;**240**:117521. Available from: https://www.sciencedirect.com/science/article/pii/S0013935123023253

[65] Machado TO, Grabow J, Sayer C, de Araújo PHH, Ehrenhard ML, Wurm FR. Biopolymer-based nanocarriers for sustained release of agrochemicals: A review on materials and social science perspectives for a sustainable future of agri- and horticulture. Advances in Colloid and Interface Science [Internet]. 2022;**303**:102645. Available from: https://www.sciencedirect.com/science/article/pii/S0001868622000471

[66] Saberi Riseh R, Vatankhah M, Hassanisaadi M, Kennedy JF. Increasing the efficiency of agricultural fertilizers using cellulose nanofibrils: A review. Carbohydrate Polymers. [Internet]. 2023;**321**:121313. Available from: https://www.sciencedirect.com/science/article/pii/S0144861723007786

[67] Dutta S, Pal S, Panwar P, Sharma RK, Bhutia PL. Biopolymeric nanocarriers for nutrient delivery and crop biofortification. ACS Omega [Internet]. 2022;7(30):25909-25920. Available from: https://www.ncbi.nlm.nih.gov/pmc/articles/PMC9352165/

[68] Júnior A, Mulinari J, De Oliveira C, Reichart F. Nanofertilizers: An overview. In: Ciência, Tecnologia e Inovação: do campo à mesa [Internet]. Instituto Internacional Despertando Vocações; 2020. DOI: 10.31692/ICIAGRO.2020.0041 [Accessed: Jan 11, 2024]

[69] Sathiyabama M. Biopolymeric nanoparticles as a nanocide for

crop protection. Nanoscience for Sustainable Agriculture. Springer [Internet]; 2019. pp. 139-152. Available from: https://link.springer.com/chapter/10.1007/978-3-319-97852-9_6

[70] Vinzant K, Rashid M, Khodakovskaya MV. Frontiers. Advanced applications of sustainable and biological nano-polymers in agricultural production. Frontiers in Plant Science [Internet]. 2022;**13**:1081165. DOI: 10.3389%2Ffpls.2022.1081165 [Accessed: Jan 11, 2024]

[71] Kondoyanni M, Loukatos D, Maraveas C, Drosos C, Arvanitis KG. Gels – Free full-text – Insights into the potential of biopolymeric aerogels as an advanced soil-fertilizer delivery systems [Internet]. 2022:**7**(2):69. DOI: 10.3390/biomimetics7020069 [Accessed: Jan 11, 2024]

[72] Gul MZ, Beedu SR. Integrated approaches to agri-nanotechnology: Applications, challenges, and future perspectives. In: Dar GH, Bhat RA, Mehmood MA, editors. Microbiomes for the Management of Agricultural Sustainability. [Internet]. Cham: Springer Nature Switzerland; 2023. pp. 1-28. DOI: 10.1007/978-3-031-32967-8_1 [Accessed: Jan 9, 2024]

[73] Bekele MT. Electrochemical sensor based on polyaniline supported CdS/CeO2/Ag3PO4 nanocomposite for malathion detection. Sensors International. [Internet]. 2024;**5**:100251. Available from: https://www.sciencedirect.com/science/article/pii/S2666351123000256

[74] Nasr N, Shafi M, Zhao T, Ali R, Ahmad I, Khan M, et al. A two-fold SPR-SERS sensor utilizing gold nanoparticles and graphene thin membrane as a spacer in a 3D composite structure. Spectrochimica Acta Part A: Molecular and Biomolecular Spectroscopy. [Internet]. 2024;**304**:123331. Available from: https://www.sciencedirect.com/science/article/pii/S1386142523010168

[75] Willner MR, Vikesland PJ. Nanomaterial enabled sensors for environmental contaminants. Journal of Nanobiotechnology [Internet]. 2018;**16**(1):95. DOI: 10.1186/s12951-018-0419-1

[76] Kim DY, Yang T, Srivastava P, Nile SH, Seth CS, Jadhav U, et al. Alginic acid-functionalized silver nanoparticles: A rapid monitoring tool for detecting the technology-critical element tellurium. Journal of Hazardous Materials [Internet]. 2024;**465**:133161. Available from: https://www.sciencedirect.com/science/article/pii/S0304389423024457

[77] Qu C, Sun W, Hu D, Yang C, Zhao T, Wang X, et al. Semi-embedded flexible multifunctional sensor for on-site continuous monitoring of plant microclimate. Computers and Electronics in Agriculture [Internet]. 2024;**216**:108521. Available from: https://www.sciencedirect.com/science/article/pii/S0168169923009092

[78] Chhipa H. Chapter 6 - Applications of nanotechnology in agriculture. In: Gurtler V, Ball AS, Soni S, editors. Methods in Microbiology [Internet], Nanotechnology. Vol. 46. Academic Press; 2019. pp. 115-142. Available from: https://www.sciencedirect.com/science/article/pii/S0580951719300029 [Accessed: Jan 10, 2024]

[79] Corradini E, de Moura MR, Mattoso LHC. A preliminary study of the incorparation of NPK fertilizer into chitosan nanoparticles. Express Polymer Letters. 2010;**4**(8):509-515

[80] Antonacci A, Arduini F, Moscone D, Palleschi G, Scognamiglio V. Nanostructured (bio)sensors for smart

agriculture. TrAC Trends in Analytical Chemistry [Internet]. 2018;**98**:95-103. Available from: https://www.sciencedirect.com/science/article/pii/S016599361730273X

[81] Singh K, Pombeiro AJL, Garazade IM, Sun Q, Mohan B. Plasmon-enhanced fluorescence sensors for food and environmental samples monitoring. Microchemical Journal [Internet]. 2024;**197**:109888. Available from: https://www.sciencedirect.com/science/article/pii/S0026265X23015072

[82] Baruah S, Dutta J. Nanotechnology applications in pollution sensing and degradation in agriculture: A review. Environmental Chemistry Letters [Internet]. 2009;7(3):191-204. DOI: 10.1007/s10311-009-0228-8

[83] Srivastava AK, Dev A, Karmakar S. Nanosensors and nanobiosensors in food and agriculture. Environmental Chemistry Letters [Internet]. 2018;**16**(1):161-182. DOI: 10.1007/s10311-017-0674-7

[84] Yao L, Xu M, Liu Y, Niu R, Wu X, Song Y. Estimating of heavy metal concentration in agricultural soils from hyperspectral satellite sensor imagery: Considering the sources and migration pathways of pollutants. Ecological Indicators [Internet]. 2024;**158**:111416. Available from: https://www.sciencedirect.com/science/article/pii/S1470160X23015583

[85] Riu J, Maroto A, Rius FX. Nanosensors in environmental analysis. Talanta [Internet]. 2006;**69**(2):288-301. Available from: https://www.sciencedirect.com/science/article/pii/S0039914005006570

[86] Abedi-Firoozjah R, Ebdali H, Soltani M, Abdolahi-Fard P, Heydari M, Assadpour E, et al. Nanomaterial-based sensors for the detection of pathogens and microbial toxins in the food industry; a review on recent progress. Coordination Chemistry Reviews. [Internet]. 2024;**500**:215545. Available from: https://www.sciencedirect.com/science/article/pii/S0010854523005349

[87] Valle AL, Silva ACA, Dantas NO, Sabino-Silva R, Melo FCC, Moreira CS, et al. Application of ZnO nanocrystals as a surface-enhancer FTIR for glyphosate detection. Nanomaterials [Internet]. 2021;**11**(2):509. Available from: https://www.mdpi.com/2079-4991/11/2/509

[88] do Valle AL, ACA S, FCC M, de Fernandes LG, de Oliveira GS, Dantas NO, et al. Glyphosate: ZnO nanocrystal interaction controlled by pH changes. IEEE Sensors Journal [Internet]. 2021;**21**(18):19731-19739. Available from: https://ieeexplore.ieee.org/document/9494482

[89] Uda MNA, Gopinath SC, Hashim U, Halim NH, Parmin NA, Uda MA, et al. Silica and graphene mediate arsenic detection in mature rice grain by a newly patterned current–volt aptasensor. Scientific Reports [Internet]. 2021;**11**(1):14688. Available from: https://www.nature.com/articles/s41598-021-94145-0 [Accessed: Jan 10, 2024]

[90] Yao Y, He HX, Ya ZB, Zhang Z, Hou D, Kun ZZ. Facile fabrication of high sensitivity cellulose nanocrystals based QCM humidity sensors with asymmetric electrode structure. Sensors and Actuators B: Chemical. [Internet]. 2020;**302**:127192. Available from: https://www.sciencedirect.com/science/article/pii/S0925400519313917